Sustainability Sciences in Asia and Africa

Sustainable Agriculture and Food Security

Series Editor

Rajeev K. Varshney, Centre for Crop and Food Innovation, Food Futures Institute
Murdoch University, Perth, WA, Australia

Sustainability as a scientific endeavor requires broad understanding of the interconnections in our global environment. It is therefore positioned at the crossroads of research in natural sciences, technology and human sciences. This book series supports the global efforts by providing timely coverage of the progress, opportunities, and challenges towards achieving sustainable development goals (SDGs) in Asia and Africa. It fosters the research in transdisciplinary academic fields spanning across agriculture and global food security, health and diseases, access to clean water and affordable energy, climate resilience and effects of global climate change on biodiversity, water and land. This series is planned to narrate the stories of scientific development and progression towards SDGs from the regions of Asia and Africa. The overall aim of the series is to provide a comprehensive publication platform and act as a knowledge engine in the growth of sustainability sciences with a special focus on developing nations. The series constitute of four sub-series each focusing on a specific theme closely linked to a set of SDGs. Each sub-series is leaders and experts in agricultural sciences, environmental sciences, biomedicine and medical sciences respectively. Books in this series are useful for environmentalists, biotechnologists, plant scientists, medical practitioners, conservation biologists, policy makers, NGOs and scientific bodies working in the areas of sustainability sciences in the developing regions.

This book series support the global efforts towards sustainability by providing timely coverage of the progress, opportunities, and challenges of sustainable food production and consumption in Asia and Africa. The series narrates the success stories and research endeavors from the regions of Africa and Asia on issues relating to SDG 2: Zero hunger. It fosters the research in transdisciplinary academic fields spanning across sustainable agriculture systems and practices, post- harvest and food supply chains. It will also focus on breeding programs for resilient crops, efficiency in crop cycle, various factors of food security, as well as improving nutrition and curbing hunger and malnutrition. The focus of the series is to provide a comprehensive publication platform and act as a knowledge engine in the growth of sustainability sciences with a special focus on developing nations. The series will publish mainly edited volumes but some authored volumes. These volumes will have chapters from eminent personalities in their area of research from different parts of the world.

Saif Uddin • Mohammad Zaman
Editors

Micro- and Nano-plastics in Soil and Crop Systems

Springer

Editors
Saif Uddin
Environment and Life Sciences
Research Centre
Kuwait Institute for Scientific Research
Safat, Kuwait

Mohammad Zaman
Soil and Water Management and Crop
Nutrition Section
International Atomic Energy Agency
Vienna, Austria

ISSN 2730-6771 ISSN 2730-678X (electronic)
Sustainability Sciences in Asia and Africa
ISSN 2730-6798 ISSN 2730-6801 (electronic)
Sustainable Agriculture and Food Security
ISBN 978-981-95-2739-7 ISBN 978-981-95-2740-3 (eBook)
https://doi.org/10.1007/978-981-95-2740-3

This Springer imprint is published by the registered company Springer Nature Singapore Pte Ltd.
The registered company address is: 152 Beach Road, #21-01/04 Gateway East, Singapore 189721, Singapore

Foreword

Micro- and nanoplastics (MNPs) are emerging contaminants of global concern. While much attention has been devoted to plastic pollution in aquatic ecosystems, the equally pressing issue of MNPs in terrestrial environments has only recently begun to receive the attention it warrants. The soil—a vital component of the Earth's ecosystem—serves as the substrate for most food production, water filtration, and carbon storage. Yet, it is increasingly compromised by the pervasive infiltration of plastic particles, ranging from macroplastics to nanoplastics. This book, *Micro- and Nanoplastics in Soil and Crop Systems*, delves into the multifaceted dimensions of plastic pollution, offering a comprehensive exploration of the interactions between plastics and terrestrial ecosystems.

The increasing population and demand for agricultural productivity and increased utilization of sewage sludge for soil amendment, use of plastic mulch films, greenhouse covers, irrigation pipes, and use of wastewater for irrigation have inadvertently introduced MNPs into soil and crop systems. Once in the soil, MNPs interact with biological, chemical, and physical processes, often altering the soil's structure, fertility, and microbial composition. Even more concerning is the potential for these particles to be absorbed by crops, leading to their infiltration into the food chain and posing risks to human and animal health.

This book addresses these critical concerns, underpinning the issues with the latest research findings, methodological advancements, and critical perspectives on the subject of MNPs, aiming to provide a reference for researchers, policymakers, and practitioners. The chapters within reflect the collaborative efforts of experts from diverse disciplines, all striving to deepen our understanding of the occurrence, behavior, and impacts of MNPs in soil and crop systems.

The Kuwait Institute for Scientific Research is deeply committed to working on cutting-edge science. I am proud that scientists at the Environment and Life Sciences Research Centre (ELSRC) are the first ones to have developed a reference material for microplastics in soil and beach sand that is shared with the International Atomic Energy Agency's (IAEA) member states. The researchers at ELSRC are working on radiolabeling the MNPs, enabling researchers to trace their pathways in the environment, study their interactions with soil and crops, and develop innovative solutions for mitigating their impacts. Kuwait Institute for Scientific Research has been the IAEA Collaborating Centre since 2019 that includes microplastics research and contributes to the Joint FAO-IAEA global effort to safeguard soil health and ensure food security in the face of emerging contaminants.

This book also reflects the importance of interdisciplinary collaboration in tackling complex environmental challenges. The integration of knowledge from soil science, agronomy, chemistry, toxicology, and environmental policy is essential for developing holistic solutions to the MNP problem. The contributors to this volume bring diverse perspectives and expertise, demonstrating the power of collective action in advancing scientific understanding and informing decision-making.

As the Director General of the Kuwait Institute for Scientific Research, it is my privilege to present this book to the global scientific community. I commend the authors for their dedication and rigorous research, which have culminated in a resource that will undoubtedly be useful to the scientific community. I hope that this book will not only enhance our understanding of MNPs in soil and crop systems but also inspire collaborative efforts to address this pressing issue.

The presence of MNPs in terrestrial ecosystems is a stark reminder of the unintended consequences of our reliance on plastic materials. However, it is also an opportunity to rethink our relationship with plastics and develop strategies to mitigate their impacts. By advancing research, fostering innovation, and promoting sustainable practices, we can work toward a future where soil—the cornerstone of life on Earth—remains healthy and productive for generations to come.

Director General
Kuwait Institute for Scientific Research
Safat, Kuwait

Faisal Al-Humaidan

Contents

Editors and Contributors

About the Editors

Saif Uddin is a Senior Research Scientist in the Environment and Life Sciences Research Centre at the Kuwait Institute for Scientific Research, Kuwait. He is a distinguished environmental scientist with over two decades of research leadership at KISR. His work focuses on the sources, pathways, and ecological impacts of pollutants—including microplastics, heavy metals, and radionuclides—across terrestrial and marine systems. He has pioneered studies on microplastic contamination in arid soils, wastewater systems, and agricultural matrices. He has been the lead scientist in various projects of national and international importance, including environmental impact assessments of Boubiyan Island development, Mubarak Al-Kabeer Port, Al-Khairan coastal development, Sheikh Jaber Causeway, Clean Fuel Project of the Kuwait National Petroleum Company, and NUTEC-Plastics project of the International Atomic Energy Agency, to name a few. He has led over 90 projects resulting in over 30 million US dollars of research funding. He has been at the forefront of environmental research at KISR and has developed the Environmental Impact Assessment, Radioecology, and Microplastics Research laboratories. His editorial expertise includes an Assistant Editor role with the *Journal of Environmental Exposure Assessment and Sustainability*. His interdisciplinary background in environmental assessment, remote sensing, and arid-land sustainability positions him uniquely to address the complex interactions of microplastics in soil-crop systems under changing climatic conditions. He has an H-index of 31, i-Index of 76, and 2985 citations as per Google Scholar; he has authored over 20 peer-reviewed publications on microplastics, including soil/crop impacts. He has two US patents to his credit one for sample collection at the air-sea neustonic layer and the other for preparation of sources for radionuclide determination. He has been doing extensive work in arid ecosystems (Kuwait, Arabian Gulf), offering insights into plastic pollution in vulnerable agricultural regions and looking at the long-term environmental implications of sludge usage in terrestrial and marine environments.

Mohammad Zaman is working as a Section Head of the Soil and Water Management and Crop Nutrition Section of the Joint FAO-IAEA Centre, working on developing climate-smart agricultural practices for enhancing food production and minimizing land degradation and environmental sustainability using nuclear and related techniques. Before joining the SWMCN Section, he worked for 22 years in GHG mitigation, enhancing soil fertility and integrated plant nutrient management at different research, academic, commercial, and international organizations in both developing and developed countries. He is the author and co-author of 3 books, several book chapters, and over 65 research publications in refereed journals. He has won several awards for his professional achievements, including the IAEA Superior Achievement Award in 2018, the IAEA Best Technical Cooperation Award in 2018, and the Lincoln University, Canterbury, New Zealand Doctorate Scholarship in 1995.

Contributors

Yasmeen Al-Babtain Environment and Life Sciences Research Centre, Kuwait Institute for Scientific Research, Safat, Kuwait

Shua'a Al-Ruwayeh Environment and Life Sciences Research Centre, Kuwait Institute for Scientific Research, Safat, Kuwait

Ahmad Al-Shemmari Environment and Life Sciences Research Centre, Kuwait Institute for Scientific Research, Safat, Kuwait

Maha Al-Sinan Environment and Life Sciences Research Centre, Kuwait Institute for Scientific Research, Safat, Kuwait

Antony Andrady Department of Chemical and Biomolecular Engineering, North Carolina State University, Raleigh, NC, USA

Montaha Behbehani Environment and Life Sciences Research Center, Kuwait Institute for Scientific Research, Safat, Kuwait

Nazima Habibi Environment and Life Sciences Research Center, Kuwait Institute for Scientific Research, Safat, Kuwait

Jin-Yong Lee Department of Geology, Kangwon National University, Chuncheon, Republic of Korea

Talat Saeed Environment and Life Sciences Research Centre, Kuwait Institute for Scientific Research, Safat, Kuwait

Saif Uddin Environment and Life Sciences Research Center, Kuwait Institute for Scientific Research, Safat, Kuwait

Mohammad Zaman Joint FAO-IAEA Centre of Nuclear Techniques in Food and Agriculture, Department of Nuclear Sciences and Applications, International Atomic Energy Agency, Vienna, Austria

Chapter 1
Microplastics in Soil: Inventories, Effect and Environmental Risks

Saif Uddin , **Nazima Habibi, and Montaha Behbehani**

Abstract The omnipresence of microplastics (MPs) in the environment has garnered immense scientific and public interest. Recent studies indicate that these contaminants are not restricted to the aquatic environment but probably are more prevalent in terrestrial systems, including agricultural soils. Soils are vulnerable to MP contamination through multiple pathways, including the disintegration of plastic mulches, sewage sludge for soil amendment, irrigation with wastewater, and atmospheric deposition. We present a comprehensive synthesis of the current literature on microplastics in soil systems, focusing on their occurrence, impacts on soil properties, including microbial assemblages, and potential to affect biogeochemical cycling. A systematic literature search was conducted using major scientific databases, and over 150 articles published between 2010 and 2024—including review papers from leading Elsevier and Springer journals—were evaluated. The findings reveal that microplastics can alter soil physical and chemical properties and affect the rhizosphere's nutrient cycling and microbiota. It also shows the lack of standardisation among the methodologies and often lack quality control in reported data. The current research provides insights into microplastic-soil interaction, but significant gaps remain regarding the long-term implications of microplastic accumulation in soils. Future studies should address standardised detection methods, elucidate crop uptake mechanisms, and assess the impact on food safety and human health. The chapter underscores the urgent need for harmonising methodologies and undertaking simulation modelling to understand the long-term effects of microplastics in soil and to design policies to mitigate MP contamination in soil systems.

Keywords Microplastics · Crop systems · Agroecosystems · Soil contamination · Food safety · Environmental risk

S. Uddin (✉) · N. Habibi · M. Behbehani
Environment and Life Sciences Research Centre, Kuwait Institute for Scientific Research, Jamal Abdul Nasser Street, Safat, Kuwait
e-mail: sdin@kisr.edu.kw

S. Uddin, M. Zaman (eds.), *Micro- and Nano-plastics in Soil and Crop Systems*, Sustainability Sciences in Asia and Africa, https://doi.org/10.1007/978-981-95-2740-3_1

1.1 Introduction

The proliferation of plastic production over the past several decades has led to widespread environmental contamination by plastic debris, with microplastics (plastic particles 0.001 and 5 mm in size) emerging as ubiquitous contaminants in various ecosystems (Jambeck 2015; Andrady 2011, 2017; Uddin et al. 2022a, b, 2023). While early research predominantly focused on marine systems, an increasing number of studies now document MPs' occurrence and potential impacts in terrestrial environments (Sajjad et al. 2022; Rillig and Lehmann 2020; Uddin et al. 2020a, 2022c, 2024; Naji et al. 2021; Jinadasa et al. 2023; Zaman and Uddin 2024; Zhang et al. 2021a). Soil is particularly vulnerable due to the application of plastic mulches (Steinmetz et al. 2016; Serrano-Ruiz et al. 2021; Huang et al. 2020; Brodhagen et al. 2015), irrigation with wastewater containing MPs (Uddin et al. 2020a, 2022c; Naji et al. 2021), atmospheric deposition (Habibi et al. 2022), and sewage sludge application for soil amendment and as fertiliser (Zaman and Uddin 2024).

1.1.1 Background and Scope

With an increasing global population, food security has emerged as a significant issue in recent times. The United Nations has included it in the Sustainable Development Goals (SDGs), i.e. SDG 2: Zero Hunger, which primarily focuses on ending hunger, achieving food security, improving nutrition, and promoting sustainable agriculture. Most countries have intensified agriculture to meet the increasing food demand (Gomiero 2016; Godfray et al. 2010). Intensive agriculture has doubled crop production (Tilman et al. 2002; Ray et al. 2012) by introducing new resilient high-yielding varieties, extensive irrigation, fertilisers (synthetic nitrogen, phosphorous), pesticides, herbicides, and soil amendments (Tilman et al. 2001). Consequent to these pressures of increasing productivity, large-scale soil degradation is reported by the United Nations Convention to Combat Desertification (UNCCD) (UNCCD 2023). The UNCCD stated that over 100 million hectares of healthy and productive land are lost yearly based on the reporting figures from 126 countries across regions, underscoring the need to address the issue of soil degradation on priority. One of the most prevalent ways is the addition of sludge to the soil to improve its nutrient content, organic matter and water retention capacity (Zaman and Uddin 2024). However, with the addition of sludge comes other pollutants like heavy metals, persistent organic pollutants, and microplastics (Uddin et al. 2024).

Another fact to consider is that agricultural land is one of the largest terrestrial biomes on the planet (Gomiero 2016). The intensification of agriculture has exerted massive pressure on the planet's water resources, and almost 70% of the water from underground aquifers and surface water bodies is used in agriculture (FAO 2021; Al-Murad et al. 2017). To overcome the issue of water scarcity, both treated and untreated wastewater has been utilised in agriculture, introducing gigantic quantities of microplastics (Uddin et al. 2020a, b, 2022c; Naji et al. 2021) and pharmaceuticals (Gevao et al. 2022) into the soils.

Agricultural practices have evolved rapidly over recent decades, with the adoption of greenhouses for climate-smart agriculture, plastic mulches to decrease water losses via evaporation and increase the temperature of soil (Kader et al. 2019; Li et al. 2022a), and drip irrigation for optimising irrigation water use to overcome climatic stressors (Wang et al. 2024). However, introducing plastics into the environment eventually degrades, resulting in gigantic microplastics and even nanoplastics quantities in different environmental matrixes, including soil (Uddin et al. 2022a, 2023; Jinadasa et al. 2023; Moller et al. 1989; Lee et al. 2024). The issue was exacerbated further when biodegradable bags and plastic were introduced since most of them were not fully biodegradable; instead, they broke into tiny particles, creating a more serious environmental issue in the long term.

The reports and claims of microplastics altering soil properties include an exhaustive list of physico-chemical properties from changes in soil texture, structure, aggregate stability, porosity, permeability, infiltration rate, bulk density, saturated hydraulic conductivity, water holding capacity, electrical conductivity, change in soil pH, nutrient cycling, reduction in soil fertility, to name a few. Besides, the MP also act as a vector transporting hydrophobic contaminants and, at the same time, can also leach loosely bounded contaminants due to degradation. There is emerging evidence of microplastic translocation into the plants and its effect on the microbiota in the rhizosphere.

1.2 Materials and Methods

1.2.1 Literature Search Strategy

A systematic literature search using the keywords microplastics in soil, plastic mulching, and microplastics in agricultural soils was performed in Web of Science, Scopus, PubMed, and Google Scholar. The literature search was restricted to peer-reviewed articles published in English between 2010 and 2024. A total of 158 articles were screened; the review articles with similar data were excluded for redundancy.

1.2.2 Inclusion and Exclusion Criteria

Articles were included if they met the following criteria:

- Were published in peer-reviewed journals
- Focused on microplastics in terrestrial settings.
- Addressed microplastic occurrence, transport, or effects in soil systems.

- Provided experimental, observational, or modelling data.
- Included sufficient methodological detail regarding MP detection and quantification.

Studies were excluded if they:

- Concentrated solely on aquatic environments or waste (sludge) without discussion of terrestrial transfer.
- Lacked sufficient methodological or statistical rigour.
- Units are not reported clearly.
- Data quality objectives are not met.

1.2.3 Data Extraction and Synthesis

For each article, key data were extracted, including:

- Study location and soil characteristics
- MP characteristics: size range, polymer types and concentrations
- Main findings: inventories and effects on soil properties

Extracted data were compiled into summary tables (see Tables 1.1 and 1.2) to facilitate study comparisons. A narrative synthesis was then developed, highlighting trends, discrepancies, and key research gaps.

1.3 Results

The literature search revealed several thematic areas where microplastics interact with soil. The findings can be broadly categorised into (1) occurrence and distribution, (2) impacts on soil properties, and (3) impact on microbial population in soil.

1.3.1 Occurrence and Distribution of Microplastics in Soil

Recent studies have documented the presence of microplastics in agricultural soils across diverse regions, including Europe, Asia, and North America (Zaman and Uddin 2024; Uddin et al. 2024; Nizzetto et al. 2016a, b; Hurley and Nizzetto 2018; Ng et al. 2018; Hanif et al. 2024; de Souza Machado et al. 2018; Zhang et al. 2020a). The main contributors of microplastics in agricultural soils are degraded plastic mulches, treated wastewater, and sewage sludge for soil amendments, besides airborne MPs, micro rubber from tyres (Smyth et al. 2025), degraded drip lines, plastic bags and others. Microplastic concentrations in agricultural soils vary significantly

Table 1.1 Microplastic presence in soils

Country/ Region	Matrix	Inventory	Size of MPs	Polymer types	Findings	References
Australia	Agroecosystems	2800 to 19,000 tons of MP/year; 2.3 and 15.8 tons MP/ha over the lifetime in the top 75–100 mm of soil				Nizzetto et al. (2016a), Nizzetto et al. (2016b)
China	Agriculture	0.56 million tonnes			Mulch covering some 20 million hectares of farmland	Espi (2006), Liu et al. (2014)
Nanjing and Wusi, Jiangsu, China	Fishpond bank soil	207–340 MP/kg				Li et al. (2019)
	Bamboo forest soil	~200 MP/kg				
	Green ChoiHouse soil	~400 MP/kg				
	Vegetable farm soil	427–573 MP/kg				
	Tea garden soil	407–740 MP/kg				
Shanghai, China	Urban soil	78.0 ± 12.9 MP/kg				Liu et al. (2018)
Shangsuan, Southern China	Cropland 1971 m above sea level	410–610 MP/kg*; 800–1080 MP/kg†; 9560–11,170 MP/kg‡	9780–11,130 MP/kg[a]; 50–70 MP/kg[b]; 410–520 MP/kg[c]; 500–1060 MP/kg[d]		Each site was 30 m x 5 m from where six random samples were collected	Zhang and Liu (2018)

(continued)

Table 1.1 (continued)

Country/ Region	Matrix	Inventory	Size of MPs	Polymer types	Findings	References
Xinjiang, China	Agricultural soils	0 to 52 kg/ha (mean 121.5 kg/ha)			Mulch reside	Zhang et al. (2016)
United States of America	Municipal soil waste	0.5–0.6% having size less than 2000 μm w/w	420 - 25,000 μm	PET, polyester, polystyrene and foam	MPs constitute 5% w/w of the MSW	
Anle, Southern China	Cropland 1908 m above sea level	690–1110 MP/kg*; 2040–2770 MP/kg†; 22,320–25,960 MP/kg‡	23,000–26,940 MP/kg[a]; 50–60 MP/kg[b]; 640–1290 MP/kg[c]; 1360–1570 MP/kg[d]			Zhang and Liu (2018)
Dunshang, Southern China	Cropland 1894 m above sea level	470–530 MP/kg*; 1490 MP/kg†;10,900–13,430 MP/kg‡	12,210–14,680 MP/kg[a]; 10 MP/kg[b]; 460–550 MP/kg[c]; 170–230 MP/kg[d]			
Dagoujian, Southern China	Cropland 1890 m above sea level	1560–1630 MP/kg*; 4060–4080 MP/kg†; 20,290–20,530 MP/kg‡	24,850–24,920 MP/kg[a]; 10–20 MP/kg[b]; 570–650 MP/kg[c]; 480–650 MP/kg[d]			
Buffer zone, Chai River Valley, southern China	Forest 1890 m above sea level	1010–1280 MP/kg*; 2580–3370 MP/kg†; 9230–11,230 MP/kg‡	11,660–13,990 MP/kg[a]; 10–50 MP/kg[b]; 670–1110 MP/kg[c]; 450–780 MP/kg[d]			
Hungary	Arable land Greenhouse	225 ± 61.69 MP/kg		PE, PVC, and PP	Tomato plantation	Sa'adu and Farsang (2022)
Shenyang Agricultural University, China	Arable land	2899 and 8885 MP/kg		PE, PP, and PET	Mulching experiment since 1987 as source and maize	Li et al. (2022b)

China	Upland Mulched soil	754 ± 477 MP/kg		PP, PE,PET, PES		Zhang et al. (2021b)
Changsha, China	Topsoil (20 cm) of agricultural land				Pot experiment; showing effect of HPDE and polyethylene in Chinese cabbage	Yang et al. (2021a)
China	Arable land	1629.68 tonnes/year			MPs resulting from the disintegration of plastic gauze. This info is based on remote sensing observation and models.	Mo et al. (2021)
Ishikawa Prefecture, Hokuriku, Japan	Paddy land	2 to 123 microcapsules/kg (mean ± SD = 48 ± 25 pieces/kg); 6–369 mg/kg.	2–5 mm in size	PE	19 sites were surveyed for microcapsules (coated fertilisers)	Katsumi et al. (2021)
Taiwan	Tainan city agriculture farms	12–117 MP/m^2	Range of sizes from 1–5 mm in form of pellets, foam, fragments, films, fibres, microbeads and others are reported	PE, LPDE, PP and PS	About 3 times higher MPs in farms adjacent to roads compared to farms away from road	Fakour et al. (2021)
Tamil Nadu, India	Arable land	Not reported in terms of numbers. Only information was given as ratio to depths (0–10 cm was 37.97%, 11–20 cm was 35.07%, and 21–30 cm was 36.99%			Plastic mulch in fields growing tomato	Kumar and Sheela (2021)

(continued)

Table 1.1 (continued)

Country/ Region	Matrix	Inventory	Size of MPs	Polymer types	Findings	References
Yeoju City, Korea	0–5 cm top soil	Average 700 MP/kg; Roadside soils 1108 MP/kg, upland soil 3440 MP/kg; Agriculture soil 664 MP/kg		PE, PP, black styrene-butadiene rubber (SBR)		Choi et al. (2021)
Switzerland	29 Floodplain soils in Swiss Natural Reserve	90% of Swiss floodplain soils contain MPs. (0.05–0.75 mg/kg).	Dominance of particles in <500 µm source.	PP, PE, HPDE, PVC, PUR, PET, PS, ABS, PA, PC		Scheurer and Bigalke (2018)
Switzerland	Eleven agricultural sites in the Swiss Seeland with high use of mulch foil, mulch tunnels and other plastic items	0–34.6 MP/L in irrigation water	PA is the most abundant polymer in the 100–150 µm size range, and PE is most abundant in the size range of 150–300 µm	Polyamide (PA) and PE are the most common polymers.		Bigalke et al. (2022)
Germany	Test Agricultural fields in Speyer	Sludge 97.7 MP/g Sludge applied soil 14.6 MP/g Control Soil 6.4 MP/g		PE, PA, polyolefins, polyacrynotrile, chloropolymer,	Only 1.4% of load was observed in deep soil where sludge was applied	Tagg et al. (2022)
Greece	Agricultural soil (0–30 cm), Ilia County	301 ± 140 MP/kg in fields planted with watermelons; 69 ± 38 MP/kg in fields with tomato plantation	20–500 µm (16 ± 6.8 MP/kg for watermelon-planted soils) and 33 ± 22 MP/kg for tomato-planted soils. 0.5–5 mm, the abundance was 286 ± 138 MP/kg for watermelon-planted soil and 36 ± 23 MP/kg for tomato-planted soil	PE and Black mulch film	Watermelon is planted twice, so input is twice; that might explain the higher MP load in watermelon planted soils	Isari et al. (2021)

Spain	Murcia (loamy soil)	0–10 cm depth 2302 ± 937 MP/kg; 10–30 cm depth 2179 ± 1069 MP/kg	30 μm–2 mm	LPDE	Plastic mulch application for 12 years of different types, including 8 LDPE+ 6 PAC + 1 biodegradable	van Schothorst et al. (2021)
Netherlands	Agricultural soils	888 ± 500 MP/kg	30 μm–2 mm	PE, PP	10 tones/ha/yr compost applied for seven years	van Schothorst et al. (2021)
Switzerland	Arable land	22.4 ± 3.3 tons/year			Organic compost	Kawecki et al. (2021)
Tanzania	Irrigated land	0.21–1.50 MP/g		PET, HDLE,LDE, PS		Kundu et al. (2022)
Mauritius	Arable land	320.0 ± 112.2–420.0 ± 244.0 MP/kg		PP and PA		Ragoobur et al. (2021)
Tunisia	Irrigated soil with treated wastewater	13.21 ± 0.89 to 852.24 ± 124.2 MP/kg		PEVA, PE, PBAT, PP		Boughattas et al. (2021)
Chile	Agricultural fields	180 ± 266 to 306 ± 360 MP/kg		PE, PP, PS	Sewage sludge application	Corradini et al. (2019)
Ontario, Canada	Arable land	Fields 1, 2 and 3 prior to the application of biosolids 18, 187 and 541 MP/kg, respectively, and post-application of biosolids in fields 1,2 and 3 were 25, 130 and 298 MP/kg		PE,PP,PS,PMMA, PA, Polyester, Nylon, Paint, EVA, EPDM	Field 1 (10.24 ha) Field 2 (26.1 ha) and Field 3 (34.4 ha) were treated with biosolids	Crossman et al. (2020)
Mexico	Arable land	400–2000 MP/kg		LPDE	Mulch	Hernández-Gutiérrez et al. (2021)

(continued)

Table 1.1 (continued)

Country/ Region	Matrix	Inventory	Size of MPs	Polymer types	Findings	References
Shaanxi Province, China	Agricultural soils	1430–3410 MP/kg	0–0.49 mm were the predominant type. Fibres were the most dominant shape	PS, PE, PP, HDPE, PVC, PET		Ding et al. (2020a)
Northeast part of the Tibetan Plateau, China	Soil	20 to 110 MP/kg	Films were the dominant shape. 88.34% of MPs were <500 μm	PP, PE		Feng et al. (2020)
Schleswig-Holstein, Northern Germany	Farmland (15 locations)	379 MP were identified in ~84 kg of dry soil. Particle abundances in the sampling units ranged from 0 to 217.8 MP/kg soil with a mean abundance of 3.7 ± 11.9 MP/kg per site		PE, PP, PP, Nylon	About 34% of sites contained MPs	Harms et al. (2021)u
Xinjiang Uygur Autonomous Region, China	Agricultural soils	80.3 ± 49.3 MP/kg in filed where mulching was used for 5 years; 308 ± 138.1 MP/kg in fields mulched for 15 years; 1075.6 ± 346.8 MP/kg soil in fields where mulching was practised for 24 years	Films	PE mulch films	394 sampling sites in 19 provinces where plastic mulches are used 0–40 cm depth were samples in 1 m^2 sires	Huang et al. (2020)

Middle Franconia, southeast Germany	An agricultural farmland (0.5 ha)	16 MPs of 1–5 mm In ~50 kg soil. 0–1.25 MP/kg with a mean abundance of 0.34 ± 0.36 MP/kg. No Plastics in 5 out of 14 samples	The total weight MPs found was 0.005 grams. Fragments and films were the most common shapes, 2 were fibres. 62.5% MPs were white. Others were transparent, blue, or green colour. No particles below 2 mm in length	10 PE, 4 PP and 2 PS particles	No plastic mulch was used, and the farmland was regularly ploughed at 20 to 30 cm depth.	Piehl et al. (2018)
Shouguang, Shandong Province and Xuzhou, Jiangsu Province, China.	Greenhouse soil	MP content in Shouguang was 380–1093 MP/kg in the control site; 1000–3786 MP/kg in the cultivated soil. In Xuzhou, 390–1000 MP/kg is in the control site, and 1300–3400 MP/kg is in the cultivated soil	Dominant size range 20–100 μm	PA, PE	A total of 34 surface soil samples (0–10 cm)	Li et al. (2021a)
Yingtan, Jiangxi province, southeast China	Experimental plots filled with Quaternary red clay	Control plots 16.4 ± 2.7 MP/kg; plots amended with pig manure for 22 years 43.8 ± 16.2 MP/kg				Yang et al. (2021b)
Murcia region, Southeast Spain	Vegetable fields (0–10 cm) where plastic mulch was used for at least 10 years	Average concentration 2116 ± 1024 MP/kg of dry soil	30–2000 μm			Beriot et al. (2021)

(continued)

Table 1.1 (continued)

Country/ Region	Matrix	Inventory	Size of MPs	Polymer types	Findings	References
Yong-In Province, mid-western Korea	Agricultural farms where soil mulching, greenhouses, and rice-cultivation paddies	10–7630 MP/kg		PE, PP, and PET		Kim et al. (2021)
Paotai Town, Shihezi City, Xinjiang, China	Agricultural soils	40.35 mg/kg; 420–1290 MP/kg	900–2000 μm. However, the most abundant were 300–800 μm in soil mulched for 30 years, i.e. small MP as the mulching period increased	PE	Mulch films used for over 30 years	Li et al. (2020)
Wutong Village and Shihezi, China	Agriculture farms intensity	0–800 MP/kg in 0–10 cm; 0–200 MP/kg in 10–20 cm; 0–2200 MP/kg in 20–30 cm	Wutong MPs in field that has had intermittent mulching for 30 years was 2200 MP/kg; In Shihezi was 900 MP/kg	LPDE	Farms with continuous and intermittent mulching, small and large size farms and low and high-intensity tillage	Meng et al. (2020)
Shouguang City, Shandong Province, Northern China	Agriculture soils	310–5698 MP/kg (average: 1444 ± 986 MP/kg)	Dominant size was 0.5 mm in 0–10 cm and <0.5 mm in the 10–25 cm layer of facility agricultural soils; the dominant shape was films and fragments	Polypropylene, ethylene-propylene copolymer, and polyethylene	Smaller MPs tend to migrate as they were more in number in deeper layer	Yu et al. (2021)

Hangzhou Bay Floodplain bordering Shanghai, Hangzhou and Ningbo, China	Floodplain agricultural soil	Average non-mulched 263 MP/kg; mulched soil average was 571 MP/kg	Films and fibres were dominant 1000–3000 μm in size	Polyethylene, polypropylene, polyester, rayon, acrylic and polyamide		Zhou et al. (2020)
Valencia, Spain	Agricultural soil	Average 930 ± 740 MP/kg (light density); 1100 ± 570 MP/kg (heavy density) in soils without application of sewage sludge Average 2130 ± 950 MP/kg (light density); 3060 ± 1680 MP/kg (heavy density) in soils with application of sewage sludge	50–1000 μm in size. Fragments, fibres and films	PP, PVC	Microplastics were found in 97% of the analysed samples	van den Berg et al. (2020)
Heilongjiang Province, China	Agricultural soil	0–8400 MP/kg; Average at sites vary between 50 ± 87 to 400 ± 692 MP/kg	50–100 μm and >100 μm	PE, PP		Zhang et al. (2020a)
Guangxi Province, China	Agricultural soil amended with sewage sludge	545.9 MP/kg after amendment with 30 tons/ha yearly; 87.6 MP/kg in soils after annual amendment with 15 t/ha annually; 5.0 MP/kg with no compost application				Zhang et al. (2020b)

PS Polystyrene; *PE* Polyethylene; *PP* Polypropylene; *HDPE* High-density polyethylene; *PVC* Polyvinyl chloride; *PET* Polyethylene terephthalate

[a]Fibres

[b]Strings

[c]Films

[d]Fragments

*>5 – 1 mm size fraction

†1–0.25 mm size fraction

‡0.25–0.05 mm size fraction

Table 1.2 Microplastics effects on soil chemical properties

Location	Soil type	Microplastic type	Effects	References
Hunan (HN), China	Red soil	PE	No significant difference in soil pH where soil was fertilised; however there was a considerable pH reduction in unfertilised soil	Li et al. (2021b)
Zhejiang (ZJ), China	Paddy soil	PE	Higher pH in fertilised soil and not in unfertilised soil	
Shandong (SD), China	Fluvo-aquic soil	PE	No effect on pH	
China	Sandy loam soil in pot experiments growing six varieties of maize (*Zea mays L.*)	PLA, PE 100–154 μm size	The soil pH increased in presence of PLA and decreased in pots with PE	Wang et al. (2020)
Netherlands	Sandy agricultural soil, Unifarm, Wageningen, Netherlands	LPDE and biodegradable mulch films	An increase in pH and a decrease in electrical conductivity was observed	Qi et al. (2020)
China	Sandy loam agricultural soils with maize plant	HDPE and PLA particles of 100–154 μm size	Soil pH increased	Yang et al. (2021c)
China	Sandy loam agricultural soil	Polyethylene (PE), polystyrene (PS), polyamide (PA), polylactic acid (PLA), polybutylene succinate (PBS) and polyhydroxy butyrate (PHB)	At a concentration of 0.2% MP, the soil pH was not much affected. However, at 2% MP, the soil pH decreased in the presence of PE and PS, and increased in experimental soils with PLA and PHB. Dissolved organic carbon content increased by low doses of PLA, PE, PS, PA, and PBS. The highest DOC was with 2% PLA. NH_4^+-N content was not affected by the high-dose MPs, but decreased by the low-dose MPs, especially PLA and PHB	Feng et al. (2022)

(continued)

Table 1.2 (continued)

Location	Soil type	Microplastic type	Effects	References
Hangzhou, China	Cultivated loessial soils from Ansai County in climate-controlled chambers	PP particles <180 μm	The addition of a low level of microplastic had a minimal effect on the OC, inorganic-N, total-P, humic and fulvic acid in the DOM solution.	Liu et al. (2017)
Nanjing, Jiangsu Province, China	Surface soil (0–20 cm) from an abandoned rice paddy field	PLA powder 20–50 μm	The presence of PLA did not affect the soil microbial community. PLA could trade off the priming effect of carbon source.	Chen et al. (2020)
Xinjiang Province, China	Farmland soil	Plastic film	Soil organic matter and inorganic nitrogen were decreased. The microbes *Actinomycetes* decreased, and *Proteobacteria* increased. Over 10 years, exposure of residual plastic film in farmland decreased the soil fertility and changed the microbial community structure	Qian et al. (2018)
Dedelow, Brandenburg, Germany	Grassland dry sandy loam soil	Polyester fibres (length of 1.28 ± 0.03 mm and a diameter of 0.030 ± 0.0008 mm. (0.4 g for 100 g soil)	Microplastics increase nutrient retention by ~70%, soil aggregation by ~18%, and soil pH by ~4%. MP impacted soil enzymes, respiration and ecosystem multifunctionality	Lozano et al. (2021)
Pingdu, Qingdao, North China.	Cropped sandy loam soils (0–20 cm)	PE in size 8.68–500 μm	Increased soil Cd availability	Wang et al. (2021)

from a few particles per kilogram to over 10^5 particles per kilogram, with higher concentrations observed in intensively managed systems (Ng et al. 2018).

An enormous quantity of microplastics (<5 mm) and possibly nanoplastics (<1 μm) are generated from plastic mulching in agricultural farms due to environmental degradation from UV radiation, thermal degradation, photo-oxidation, mechanical breakdown, erosion and biological activity (e.g., earthworm ingestion) <5 mm in size (Steinmetz et al. 2016; Horton et al. 2017; Gasperi et al. 2018). Wastewater is a conspicuous source of microplastic input into soils. The soil irrigated with wastewater has higher MP concentrations, i.e. 5190 MP/kg, compared to non-irrigated soil, where an average of 2030 particles kg^{-1} (Nizzetto et al. 2016b).

Some of the thermoplastics can persist in soil for a long time undergoing fragmentation and may even become nanoplastics (Ng et al. 2018). Studies have shown that using compost can contribute significantly to microplastic soils; an estimate based on experimental results indicated an average abundance of 2400 ± 358 MP kg^{-1} (dry weight) in the compost product (Gui et al. 2021). This study stipulated that during composting, a single large expanded polystyrene (EPS), polyethene (PE), and polypropylene (PP) macroplastic could release 4–63 MPs (Gui et al. 2021). Others have also highlighted significant input of microplastics from compost and reprocessed waste (Gui et al. 2021).

Several studies have underscored the ecological impacts of microplastic addition on soil, but over 120 studies have quantified microplastics in soils (Sa'adu and Farsang 2023). Another significant issue is the discrepancies in data quality, which underscore the need for a standardised monitoring and assessment protocol. The soil microplastic data is summarised in Table 1.1.

Meanwhile, the meta-analysis of studies (see Table 1.1) confirms a significant presence of microplastics in soils globally. The data highlight significant variability in MP concentrations, polymer types, and particle sizes. The heterogeneity in methodologies used, however, renders comparison infructuous. The variation in the kind and inventories of soil microplastics has steady links to the sources, i.e., in regions where the sewage sludge is used for soil amendment or the use of compost or biochar regions depending on regional practices and environmental conditions

The annual MP input from degraded mulches and the use of wastewater for irrigation in Australia resulted in a reported accumulation of 2.3–15.8 tons ha^{-1} of microplastics in the top 10 cm of soil. The total annual microplastic input is reported to vary between 2800 and 19,000 tonnes. Several studies from China highlighted a significant microplastic concentration originating from plastic mulches in farmlands. In Shangxuan province alone, over 20 million hectares of agricultural farms use plastic mulches. The enormous concentration of 9780–11,130 MP kg^{-1} in 50–250 μm size fractions, whereas in other studies for different types of soils, including forest, greenhouse soil, and vegetable farms, the concentrations were 200–573 MP kg^{-1}. At the bank of a fishpond, the soil has a concentration of 207–340 MP kg-1. In Tea garden soil, the microplastics were 407–740 particles per kilogram, whereas in urban soils in Shanghai, microplastics are believed to result from atmospheric deposition and urban runoff. The MP in urban soils were 78.0 ± 12.9 MP kg^{-1}.

The presence of microplastics in agricultural soils in Germany that are treated with biosolids/sludge has a higher microplastic concentration of 14.6 MP g^{-1} of soil compared to agricultural soil where the sludge was not applied 6.4 MP g^{-1}. The dominant polymers in these soils were PE, PA and PS. Microplastics were also reported from greenhouses in Hungary that were planted with tomatoes; a concentration of 225 ± 61.69 MP kg^{-1} of soil was reported. The main polymers were polyethene (PE), polyvinyl chloride (PVC), and polypropylene (PP), which are essentially derived from greenhouse materials and biosolids. The floodplain soils in Switzerland were also contaminated with microplastics. About 90% of the samples had microplastics ranging between 0.05 and 0.75 mg kg^{-1} of soil. The main polymers detected were PA and PE, which are linked to irrigation water. Microplastics in Spanish agricultural soils are omnipresent, especially due to the excessive usage of sewage sludge and plastic mulch for agricultural applications. In Murcia, Spain, a top 10 cm of soil where plastic mulch was used for 12 years showed a microplastic load of 2302 ± 937 MP kg^{-1} of soil.

The United States Environmental Protection Agency (USEPA) has authorised the use of biosolids on agricultural lands, forests, rangelands, and degraded lands that are to be reclaimed. The reports that the biosolids have 0.5–0.6% MP in a size range of 420–25,000 μm raise the concern of likely physico-chemical changes in soil properties due to the presence of PET, polyester and polystyrene foams that are commonly reported. Biosolids are also applied in agricultural soils in Canada. A report from Ontario reported 18–541 MP kg^{-1} of soil before biosolids application, which changed to 25–298 MP kg^{-1} post-biosolid application. This reduction in number can be due to the downward movement of MPs in the soil profile or uptake by the microbial communities and within the plant's root systems.

In Tamil Nadu, India, the surface soil layer up to 30 cm depth has a high concentration of plastic mulch residues, with the highest concentration amounting to 37.97% in the top 10 cm. In developed nations like Japan, microcapsules are used to add fertilisers, and reports of microcapsules from paddy soils in the Ishikawa region are quite striking. An average of 48 ± 26 microcapsules per kilogram of soil have been reported. Microplastic particles of PET and PS have also been reported from irrigated soils in Tanzania, where the concentrations are as high as 0.21–1.50 MP g^{-1}.

The data availability on microplastics in soil demonstrates the wide range of concentrations that vary across several orders of magnitude within similar applications of sludge or biosolids within the soil, use of treated wastewater for irrigation and use of plastic mulches. The reasons for this deviation can be the different methodologies and protocols used. There is little information on data quality objectives, with a few exceptions. Most studies have not reported on positive controls and blanks. Several studies have reported numbers of MPs in a unit mass of soil, while others report microplastic weight in micrograms and even percentage of soil mass.

1.3.2 Effects on Soil Properties

An increasing microplastic load in soil has been demonstrated to influence its physical and chemical properties (Hasan and Tarannum 2025). Microplastics are reported to have the potential to alter soil structure, soil aggregate stability, water-holding capacities, and nutrient dynamics (de Souza Machado et al. 2018; Rillig et al. 2019; Lian et al. 2022). The microplastics are known to have the potential to act as vectors for contaminants and can desorb these contaminants along the way in the soil profile. Besides, microplastics themselves are chemically complex and can leach these chemicals into the soil as they degrade. Laboratory and field experiments have also indicated changes in soil porosity affecting the water dynamics (de Souza Machado et al. 2018). Degraded PE microplastics reduce the soil nitrogen and phosphorus levels in bulk soils (Lian et al. 2022). The changes are not limited to the soil's physical attributes, i.e. blockage of root pores, altered water availability, and chemical change (levels of organic and metal contamination) (Kumar et al. 2020; Khalid et al. 2021; Chen et al. 2024), but also affect the rhizosphere's microbial community, which can be crucial for nutrient cycling, plant growth, and health (Lian et al. 2022). Studies have reported reduced seed germination rates and lower crop yields in microplastic-contaminated soils (Chang et al. 2024).

1.3.2.1 Soil Structure and Porosity

Microplastics are reported to obliterate the soil structure, a vital element defining soil fertility and productivity. Microplastics in soils influence the movement of water, gas exchange, and root penetration. Microplastic incorporation in the soil matrix perturbs the natural aggregation process since they occur in varied shapes (fibres, fragments, granules, films) and sizes (1–5000 μm), altering the spatial arrangement of soil particles, changing the bulk density, porosity and permeability (Hasan and Tarannum 2025). A study reported that irregularly shaped microplastics can interfere with the formation of stable soil aggregates.

1.3.2.2 Water Retention and Infiltration

Microplastics in soil may significantly alter the water-holding capacity and infiltration rates. They have been observed to increase the number of macropores, which can decrease water and nutrient retention capacities (Zhang and Liu 2018). Microplastics with hydrophobic surfaces can also induce water repellency, further affecting infiltration and potentially increasing surface runoff. These hydrological alterations in soil characteristics can affect the retention of nutrients and contaminants within the soil profile.

1.3.2.3 Bulk Density and Soil Aggregation

Microplastics affect soil bulk density, which is also a measure of compaction. Since most microplastics are lower in density than mineral grains, their incorporation can decrease the overall soil density and potentially alter root penetration (Rillig and Lehmann 2020). As many agricultural fields receive a continuous input of sludge or biosolids where the microplastic abundance is high over time, soil aggregate structures can change to less stable forms. This warrants a need for long-term studies to understand how the physical integrity of soils evolves with sustained microplastic contamination.

1.3.2.4 Changes in pH and Nutrient Dynamics

Microplastics may affect the chemical equilibrium of soils by influencing pH and nutrient availability. There is a report of an increase in soil pH due to high-density polyethylene (HDPE) and polylactic acid (PLA) at 1 and 10% concentration (Wang et al. 2022). The pH also impacted soil microplastic sorption; the sorption capacity decreased as pH increased (Hüffer et al. 2018). Polystyrene microplastics have shown changes in surface sorption, hydrophobicity, and electrostatic interactions (Luo et al. 2020). Some research indicates that certain plastic additives, such as phthalates and bisphenol- A (BPA), can leach into the soil matrix, altering its chemical characteristics (de Souza Machado et al. 2018). Leaching these compounds may lead to localised changes in pH, affecting nutrient solubility and bioavailability. For example, pH shifts can alter the speciation of essential nutrients such as phosphorus and nitrogen, impacting plant uptake and microbial metabolism (Zhang and Liu 2018). Additionally, the presence of microplastics can interfere with the cation exchange capacity of soils, potentially modifying the retention and mobility of both nutrients and pollutants.

1.3.3 Sorption of Contaminants and Pollutants

Although plastics are considered stable materials, they undergo physical and chemical degradation. Once plastic enters the environment, it degrades and disintegrates into smaller sizes, and as the microplastics age, they are more likely to absorb pollutants (Ding et al. 2020a, b). Microplastics in soil are not chemically inert. Their surfaces can adsorb various contaminants, including heavy metals, pesticides, and organic pollutants. The sorptive capacity of microplastics is primarily determined by their polymer composition, surface area, and degree of weathering (Sánchez et al. 2022). Studies suggest that microplastics can act as a vector for pollutants that bind and concentrate pollutants, potentially enhancing their bioavailability to soil biota. For instance, plastic particles with high surface-area-to-volume ratios have been found to adsorb hydrophobic organic compounds, which may desorb and enter the soil solution under certain environmental conditions. This dual role—as both a sink and a source of contaminants—complicates the assessment of risks associated with microplastic pollution. The studies conducted on microplastics inducing chemical changes in soil are summarised in Table 1.2.

Some studies show that microplastics in soil perturb the soil's organic matter, which is critical for soil fertility. Microplastics also influence nutrient cycling, water retention, and the microbial community in the rhizosphere. The presence of microplastics may alter the decomposition and stabilisation processes of organic matter. The studies in Table 1.2 highlight that microplastics can hinder the absorption of organic compounds onto soil particles, altering carbon sequestration dynamics. The change often leads to less diverse microbial communities in the root zone, potentially obliterating the rate of organic matter disintegration and nutrient and metal uptakes. The microplastics perturb methane (CH_4) emissions; polyethene (PE) microplastics with hydrochar enhance methane emission and suppressed NO_2 (Han et al. 2022a), while polyethene terephthalate (PET) microplastics with biochar reduce CH_4 emission by altering redox conditions (Han et al. 2022b).

1.3.4 *Effects on Soil Microbial Communities and Ecosystem Functions*

The change in soil's physico-chemical characteristics perturbs the soil microbiome. Microplastics in the soil change the porosity and permeability, significantly affecting water absorption and retention capacities. Microplastics also affect nutrient and trace metal availability, indirectly affecting microbial community structure. The presence of polyethene and polystyrene microplastics in soil decreased the abundance of *Proteobacteria* and *Actinobacteria*, which are crucial for nutrient cycling and maintaining soil fertility (Aralappanavar et al. 2024). Microplastics create a conducive environment for microbes capable of metabolising synthetic polymers, like stress-tolerant fungi affecting nitrogen cycling (Seeley et al. 2020). At a concentration of $\leq 1\%$, microplastics have been seen to enhance nitrogen (N)-fixing and phosphorus (P)-solubilising bacteria *Burkholderiaceae* and *Xanthobacteriaceae*, increasing bioavailable N and P but inhibiting nitrifiers like ammonia-oxidising bacteria (AOB), reducing nitrification rates (Seeley et al. 2020; Zhu et al. 2022). This disruption in microbial activity elevates nitrous oxide (N_2O) emissions due to incomplete denitrification (Al Ghadban et al. 2012). Microplastics stimulate microbial respiration by increasing dissolved organic carbon and soil porosity, leading to higher CO_2 emissions (Gao et al. 2021; Zhang et al. 2022). Microplastics with co-contaminants, including heavy metals and antibiotics, exacerbate soil antimicrobial resistance and disrupt nutrient mineralisation (Zeb et al. 2022; Wang et al. 2023).

Microplastics act as vectors for organic contaminants and leach additives like phthalates and bisphenol-A into the soil, inducing toxicity on microbial species and resulting in lower microbial biomass and diversity (de Souza Machado et al. 2018) and having a cascading effect on overall soil fertility.

1.4 Mechanistic Insights and Interactions

Complex interaction occurs between microplastics and the soil matrix, bringing about physico-chemical changes. However, understanding the mechanisms underlying the physico-chemical effects of microplastics in soil requires a multidisciplinary approach. The current understanding is underpinned by results from laboratory experiments, field studies, and simulation modelling demonstrating the impact of polymer types, shape, and size on soil bulk density, porosity, permeability, efficacy as a vector of pollutants and their toxicity. Studies have also demonstrated that plastic degradation by mechanical abrasion or UV weathering enhances its ability to adsorb contaminants.

Several field studies, especially in areas that use plastic mulches or sewage sludge for soil amendment, exhibit altered physico-chemical properties compared to control (uncontaminated areas). The field experiments and mesocosms also highlight the variation in response to microplastic presence depending on the concentration of MPs, polymer type, soil type, microbial assemblages, plantation/crops and land use history.

Environmental-ecological modelling provides an edge and is beginning to integrate diverse environmental parameters to predict the long-term impacts of microplastics on soil ecosystems. Some of the modelled results suggest that even low microplastic concentrations can profoundly affect the microbes in the rhizosphere and the soil structure when exposed for a long time (decades).

Evidence has emanated on several indirect effects of microplastics in soil, such as the co-presence of microplastics and trace metals modifying the microbial community, which alters the bioavailability of metals or enhances the toxic effect on biota. There is a need to pursue research to understand the interactive and cumulative impact of multi-stressors on soil health.

1.5 Future Perspectives and Research Gaps

Limited efforts have been put into studying microplastics in the terrestrial environment, although most of the microplastics originate from land and land-based sources. The limited information on microplastic inventories in soil, quantities, types of polymers, and sizes of microplastics has made it challenging to comprehensively elucidate the effects of microplastics on soil physico-chemical properties. Without a harmonised methodology for microplastics in soil, different sampling strategies, microplastic extraction, quantification, and characterisation limit comparison across studies, emphasising the need to develop a standardised protocol.

Most of the reported studies have examined the short-term effects of microplastics in soil employing mesocosms and pot experiments. However, long-term field studies and a multi-stressor approach are essential to understand the cumulative impact of contaminants that coexist with soil microplastics and how they affect soil health.

Understanding microplastic interactions with contaminants and their mechanistic behaviour, ecological response, and bioavailability is not fully understood. This understanding is vital for assessing toxicity, risk assessment, and designing remediation strategies. The information on how soil moisture, temperature, and pH influence the carbon content and greenhouse emissions in the presence of different polymers is limited and more research is needed.

Additionally, emerging research on microplastics' biological and abiotic degradation pathways in soil offers promising avenues for understanding their persistence and transformation in soil. Studies focusing on the role of soil microbes in the biodegradation of microplastics could provide insights into natural attenuation processes and inform the development of bioremediation strategies.

Finally, looking at the complexity and interdisciplinary nature of microplastics in soil, it is essential to bring together experts in soil science, polymer chemistry, microbiology, and environmental toxicology to unravel the multifaceted impacts of microplastics in soil and support the development of comprehensive management strategies aimed at mitigating their environmental and ecological risks.

1.6 Conclusion

Microplastics are a marker horizon of the Anthropocene, with gigantic quantities added to the terrestrial environment due to the degradation of plastic mulch films, the use of sludge for soil amendment, composts and wastewater for irrigation. Consequently, microplastics are regarded as emerging contaminants with profound implications on the physico-chemical properties of soil, besides the impact on the microbiota in the rhizosphere that is critical for maintaining healthy soil. Microplastics are seen to influence the soil structure, hydrodynamics, nutrient cycling, and contaminant behaviour, posing significant challenges to the integrity and functionality of soil systems. Research conducted between 2010 and 2025 has provided valuable insights into these mechanisms, yet many uncertainties persist regarding microplastics' long-term behaviour and synergistic impacts in diverse soil environments.

The role of microplastics—as physical disruptors, chemical vectors, and substrates for microbiota (plastisphere)—highlights the complexity of their interactions with soil components. The growing volume of literature underpins the evidence indicating microplastics can disrupt soil aggregation, alter porosity and permeability, water-holding capacities, and modify the chemical equilibria of soils by leaching additives like phthalates, BPA, etc. and adsorbing metals and organic contaminants. Their presence in soil provides a unique substrate for microbes. Some eukaryotic forms prefer to colonise these polymers, affecting the soil microbial communities and key ecosystem functions, potentially affecting agricultural productivity and soil health. The potential of microplastics in creating imbalances in biogeochemical response and amplifying greenhouse gas fluxes, especially of CO_2, CH_4 and N_2O, further underscores the need for mitigation strategies to address greenhouse gas emissions and their contribution to global warming.

This synthesis presents the current knowledge on microplastics' occurrence, uptake, and impacts in soil systems. Integrating findings from diverse studies highlights the complexities and emerging trends in this field. Given the rapid expansion of plastic production and use, future research and policy efforts must prioritise the mitigation of microplastic contamination in soil systems. By advancing our understanding of the mechanisms governing microplastic interactions with soil-microbes-plants, stakeholders—from researchers to policymakers—can work together to develop sustainable agricultural practices that safeguard ecosystem health and food security.

References

Al Ghadban AN et al (2012) Denitrification potential of the Northern Arabian Gulf – an experimental study. Environ Monit Assess.184(12):7103-12 https://doi.org/10.1007/s10661-011-2483-y.
Al-Murad M et al (2017) Waterlogging in arid agriculture areas due to improper groundwater management—an example from Kuwait. Sustainability 9(11):2131
Andrady AL (2011) Microplastics in the marine environment. Mar Pollut Bull 62(8):1596–1605
Andrady AL (2017) The plastic in microplastics: a review. Mar Pollut Bull 119(1):12–22
Aralappanavar VK et al (2024) Effects of microplastics on soil microorganisms and microbial functions in nutrients and carbon cycling – a review. Sci Total Environ 924:171435
Beriot N et al (2021) Low density-microplastics detected in sheep faeces and soil: a case study from the intensive vegetable farming in Southeast Spain. Sci Total Environ 755:142653
Bigalke M et al (2022) Microplastics in agricultural drainage water: a link between terrestrial and aquatic microplastic pollution. Sci Total Environ 806:150709
Boughattas I et al (2021) Assessing the presence of microplastic particles in Tunisian agriculture soils and their potential toxicity effects using Eisenia andrei as bioindicator. Sci Total Environ 796:148959
Brodhagen M et al (2015) Biodegradable plastic agricultural mulches and key features of microbial degradation. Appl Microbiol Biotechnol 99(3):1039–1056
Chang N et al (2024) Unveiling the impacts of microplastic pollution on soil health: a comprehensive review. Sci Total Environ 951:175643
Chen H et al (2020) Mixing effect of polylactic acid microplastic and straw residue on soil property and ecological function. Chemosphere 243:125271
Chen L et al (2024) Meta-analysis of impacts of microplastics on plant heavy metal(loid) accumulation. Environ Pollut 348:123787
Choi Y et al (2021) Plastic contamination of forest, urban, and agricultural soils: a case study of Yeoju City in the Republic of Korea. J Soils Sediments 21:1–12
Corradini F et al (2019) Evidence of microplastic accumulation in agricultural soils from sewage sludge disposal. Sci Total Environ 671:411–420
Crossman J et al (2020) Transfer and transport of microplastics from biosolids to agricultural soils and the wider environment. Sci Total Environ 724:138334
de Souza Machado AA et al (2018) Microplastics as an emerging threat to terrestrial ecosystems. Glob Chang Biol 24:1405–1416
Ding L et al (2020a) The occurrence and distribution characteristics of microplastics in the agricultural soils of Shaanxi Province, in north-western China. Sci Total Environ 720:137525
Ding L et al (2020b) High temperature depended on the ageing mechanism of microplastics under different environmental conditions and its effect on the distribution of organic pollutants. Water Res 174:115634
Espi E (2006) PLastic films for agricultural applications. J Plast Film Sheet 22:85–102
Fakour H et al (2021) Quantification and analysis of microplastics in farmland soils: characterization, sources, and pathways. Agriculture 11:330
FAO (2021) The state of the world's land and water resources for food and agriculture—systems at breaking point. Synthesis report 2021, Rome. https://doi.org/10.4060/cb7654e
Feng S et al (2020) Analysis of microplastics in a remote region of the Tibetan Plateau: implications for natural environmental response to human activities. Sci Total Environ 739:140087
Feng X et al (2022) Microplastics change soil properties, heavy metal availability and bacterial community in a Pb-Zn-contaminated soil. J Hazard Mater 424:127364
Gao B et al (2021) Microplastic addition alters the microbial community structure and stimulates soil carbon dioxide emissions in vegetable-growing soil. Environ Toxicol Chem 40(2):352–365
Gasperi J et al (2018) Microplastics in air: are we breathing it in? Curr Opin Environ Sci Health 1:1–5
Gevao B et al (2022) Antibiotics in wastewater: baseline of the influent and effluent streams in Kuwait. Toxics 10(4):174

Godfray HCJ et al (2010) Food security: the challenge of feeding 9 billion people. Science 327(5967):812–818

Gomiero T (2016) Soil degradation, land scarcity and food security: reviewing a complex challenge. Sustainability 8(3):281

Gui J et al (2021) Microplastics in composting of rural domestic waste: abundance, characteristics, and release from the surface of macroplastics. Environ Pollut 274:116553

Habibi N et al (2022) Microplastics in the atmosphere: a review. J Environ Expo Assess 1(1):6. DOI: 10.20517/jeea.2021.07

Han L et al (2022a) Co-occurrence of microplastics and hydrochar stimulated the methane emission but suppressed nitrous oxide emission from a rice paddy soil. J Clean Prod 337:130504

Han L et al (2022b) Influence of polyethylene terephthalate microplastic and biochar co-existence on paddy soil bacterial community structure and greenhouse gas emission. Environ Pollut 292:118386

Hanif MN et al (2024) Impact of microplastics on soil (physical and chemical) properties, soil biological properties/soil biota, and response of plants to it: a review. Int J Environ Sci Technol 21(16):10277–10318

Harms IK et al (2021) Amount, distribution and composition of large microplastics in typical agricultural soils in Northern Germany. Sci Total Environ 758:143615

Hasan MM, Tarannum MN (2025) Adverse impacts of microplastics on soil physicochemical properties and crop health in agricultural systems. J Hazard Mater Adv 17:100528

Hernández-Gutiérrez E et al (2021) Morphospecies abundance of above-ground invertebrates in agricultural systems under glyphosate and microplastics in South-Eastern Mexico. Environments 8(11):130

Horton AA et al (2017) Microplastics in freshwater and terrestrial environments: evaluating the current understanding to identify the knowledge gaps and future research priorities. Sci Total Environ 586:127–141

Huang Y et al (2020) Agricultural plastic mulching as a source of microplastics in the terrestrial environment. Environ Pollut 260:114096

Hüffer T, Weniger A-K, Hofmann T (2018) Sorption of organic compounds by aged polystyrene microplastic particles. Environ Pollut 236:218–225

Hurley RR, Nizzetto L (2018) Fate and occurrence of micro(nano)plastics in soils: knowledge gaps and possible risks. Curr Opin Environ Sci Health 1(6–11):6

Isari EA et al (2021) Microplastics in agricultural soils: a case study in cultivation of watermelons and canning tomatoes. Water 13(16):2168

Jambeck JR et al (2015) Plastic waste inputs from land into the ocean. Science 347(6223):768–771

Jinadasa BKKK, Uddin S, Fowler SW (2023) Microplastics (MPs) in marine food chains: is it a food safety issue? In: Ozogul F (ed) Advances in food and nutrition research. Academic Press, pp 101–140. DOI: 10.1016/bs.afnr.2022.07.005

Kader MA et al (2019) Mulching as water-saving technique in dryland agriculture: review article. Bull Natl Res Cent 43(1):147

Katsumi N et al (2021) Accumulation of microcapsules derived from coated fertilizer in paddy fields. Chemosphere 267:129185

Kawecki D, Goldberg L, Nowack B (2021) Material flow analysis of plastic in organic waste in Switzerland. Soil Use Manag 37(2):277–288

Khalid N et al (2021) Interactions and effects of microplastics with heavy metals in aquatic and terrestrial environments. Environ Pollut 290:118104

Kim S-K et al (2021) Abundance and characteristics of microplastics in soils with different agricultural practices: importance of sources with internal origin and environmental fate. J Hazard Mater 403:123997

Kumar MV, Sheela AM (2021) Effect of plastic film mulching on the distribution of plastic residues in agricultural fields. Chemosphere 273:128590

Kumar M et al (2020) Microplastics as pollutants in agricultural soils. Environ Pollut 265:114980

Kundu MN, Komakech HC, Lugomela G (2022) Analysis of macro- and microplastics in riverine, riverbanks, and irrigated farms in Arusha, Tanzania. Arch Environ Contam Toxicol 82(1):142–157

Lee J-Y et al (2024) A comprehensive review of urban microplastic pollution sources, environment and human health impacts, and regulatory efforts. Sci Total Environ 946:174297

Li Q et al (2019) Separation and identification of microplastics from soil and sewage sludge. Environ Pollut 254:113076

Li W et al (2020) Microplastics in agricultural soils: extraction and characterization after different periods of polythene film mulching in an arid region. Sci Total Environ 749:141420

Li Q et al (2021a) Are microplastics correlated to phthalates in facility agriculture soil? J Hazard Mater 412:125164

Li H-Z et al (2021b) Long-term fertilization history alters effects of microplastics on soil properties, microbial communities, and functions in diverse farmland ecosystem. Environ Sci Technol 55(8):4658–4668

Li F et al (2022a) Straw mulching for enhanced water use efficiency and economic returns from soybean fields in the Loess Plateau China. Sci Rep 12(1):17111

Li S et al (2022b) Macro- and microplastic accumulation in soil after 32 years of plastic film mulching. Environ Pollut 300:118945

Lian Y et al (2022) Effects of polyethylene and polylactic acid microplastics on plant growth and bacterial community in the soil. J Hazard Mater 435:129057

Liu EK, He WQ, Yan CR (2014) 'White revolution' to 'white pollution'—agricultural plastic film mulch in China. Environ Res Lett 9(9):091001

Liu H et al (2017) Response of soil dissolved organic matter to microplastic addition in Chinese loess soil. Chemosphere 185:907–917

Liu M et al (2018) Microplastic and mesoplastic pollution in farmland soils in suburbs of Shanghai, China. Environ Pollut 242:855–862

Lozano YM et al (2021) Effects of microplastics and drought on soil ecosystem functions and multifunctionality. J Appl Ecol 58(5):988–996

Luo Y et al (2020) Distribution characteristics and mechanism of microplastics mediated by soil physicochemical properties. Sci Total Environ 726:138389

Meng F et al (2020) Effects of plastic mulching on the accumulation and distribution of macro and micro plastics in soils of two farming systems in Northwest China. PeerJ 8:e10375

Mo X et al (2021) Estimation of soil microplastic input derived from plastic gauze using a simplified model. Sci Total Environ 793:148577

Moller L et al (1989) The air pollutant 2-nitrofluorene as initiator and promoter in a liver model for chemical carcinogenesis. Carcinogenesis 10(3):435–440

Naji A et al (2021) Microplastics in wastewater outlets of Bandar Abbas city (Iran): a potential point source of microplastics into the Persian Gulf. Chemosphere 262:128039

Ng E-L et al (2018) An overview of microplastic and nanoplastic pollution in agroecosystems. Sci Total Environ 627:1377–1388

Nizzetto L, Langaas S, Futter M (2016a) Pollution: do microplastics spill on to farm soils? Nature 537(7621):488–488

Nizzetto L, Futter M, Langaas S (2016b) Are agricultural soils dumps for microplastics of urban origin? Environ Sci Technol 50(20):10777–10779

Piehl S et al (2018) Identification and quantification of macro- and microplastics on an agricultural farmland. Sci Rep 8(1):17950

Qi Y et al (2020) Effects of plastic mulch film residues on wheat rhizosphere and soil properties. J Hazard Mater 387:121711

Qian H et al (2018) Effects of soil residual plastic film on soil microbial community structure and fertility. Water Air Soil Pollut 229:1–11

Ragoobur D, Huerta-Lwanga E, Somaroo GD (2021) Microplastics in agricultural soils, wastewater effluents and sewage sludge in Mauritius. Sci Total Environ 798:149326

Ray DK et al (2012) Recent patterns of crop yield growth and stagnation. Nat Commun 3(1):1293

Rillig MC, Lehmann A (2020) Microplastic in terrestrial ecosystems. Science 368(6498):1430–1431
Rillig MC et al (2019) Microplastic effects on plants. New Phytol 223(3):1066–1070
Sa'adu I, Farsang A (2022) Greenhouse farming as a source of macroplastic and microplastics contamination in agricultural soils: a case study from Southeast-Hungary. Agrokém Talajt 71(1):43–57
Sa'adu I, Farsang A (2023) Plastic contamination in agricultural soils: a review. Environ Sci Eur 35(1):13
Sajjad M et al (2022) Microplastics in the soil environment: a critical review. Environ Technol Innov 27:102408
Sánchez A et al (2022) Dietary microplastics: occurrence, exposure and health implications. Environ Res 212:113150
Scheurer M, Bigalke M (2018) Microplastics in Swiss floodplain soils. Environ Sci Technol 52(6):3591–3598
Seeley ME et al (2020) Microplastics affect sedimentary microbial communities and nitrogen cycling. Nat Commun 11(1):2372
Serrano-Ruiz H, Martin-Closas L, Pelacho AM (2021) Biodegradable plastic mulches: impact on the agricultural biotic environment. Sci Total Environ 750:141228
Smyth K et al (2025) Pavement wear generates microplastics in stormwater runoff. J Hazard Mater 481:136495
Steinmetz Z et al (2016) Plastic mulching in agriculture. Trading short-term agronomic benefits for long-term soil degradation? Sci Total Environ 550:690–705
Tagg AS et al (2022) Agricultural application of microplastic-rich sewage sludge leads to further uncontrolled contamination. Sci Total Environ 806:150611
Tilman D et al (2001) Forecasting agriculturally driven global environmental change. Science 292(5515):281–284
Tilman D et al (2002) Agricultural sustainability and intensive production practices. Nature 418(6898):671–677
Uddin S, Fowler SW, Behbehani M (2020a) An assessment of microplastic inputs into the aquatic environment from wastewater streams. Mar Pollut Bull 160:111538
Uddin S, Fowler SW, Saeed T (2020b) Microplastic particles in the Persian/Arabian Gulf – a review on sampling and identification. Mar Pollut Bull 154:111100
Uddin S et al (2022a) Micro-nano plastic in the aquatic environment: methodological problems and challenges. Animals 12(3):297
Uddin S et al (2022b) A preliminary assessment of size-fractionated microplastics in indoor aerosol-Kuwait's baseline. Toxics 10(2):71
Uddin S et al (2022c) Microplastics in Kuwait's wastewater streams. Sustainability 14(23):15817
Uddin AS, Uddin S, Fowler SW (2023) Environmental microplastics: a significant pollutant of the Anthropocene. In: Khan A, Wang C, Asiri AM (eds) Microplastic sources, fate and solution. Springer Nature Singapore, Singapore, pp 89–105. DOI: 10.1007/978-981-99-0695-6_5
Uddin S et al (2024) Sewage sludge as soil amendment in Arid soils - a trace metal, nutrient and trace organics perspective. Emerg Contamin 11(1):100420
UNCCD (2023) At least 100 million hectares of healthy land now lost each year. United Nations Convention to Combat Desertification. https://www.unccd.int/sites/default/files/2023-10/Data%20Dashboard%20press%20release%20EN.pdf
van den Berg P et al (2020) Sewage sludge application as a vehicle for microplastics in eastern Spanish agricultural soils. Environ Pollut 261:114198
van Schothorst B et al (2021) Sources of light density microplastic related to two agricultural practices: the use of compost and plastic mulch. Environments 8(4):36
Wang X-X et al (2020) Mycorrhizal impacts on root trait plasticity of six maize varieties along a phosphorus supply gradient. Plant Soil 448:71–86
Wang F, Wang X, Song N (2021) Polyethylene microplastics increase cadmium uptake in lettuce (Lactuca sativa L.) by altering the soil microenvironment. Sci Total Environ 784:147133

Wang Y et al (2022) Investigation of microplastic pollution on paddy fields in Xiangtan City, Southern China. Environ Sci Pollut Res 29(54):81300–81307

Wang B et al (2023) Combined effects of microplastics and cadmium on the soil-plant system: phytotoxicity, Cd accumulation and microbial activity. Environ Pollut 333:121960

Wang C et al (2024) A review of the application and impact of drip irrigation under plastic Mulch in agricultural ecosystems. Agronomy 14(8):1752

Yang M et al (2021a) Influences of different source microplastics with different particle sizes and application rates on soil properties and growth of Chinese cabbage (Brassica chinensis L.). Ecotoxicol Environ Saf 222:112480

Yang J et al (2021b) Abundance and morphology of microplastics in an agricultural soil following long-term repeated application of pig manure. Environ Pollut 272:116028

Yang W et al (2021c) Effects of microplastics on plant growth and arbuscular mycorrhizal fungal communities in a soil spiked with ZnO nanoparticles. Soil Biol Biochem 155:108179

Yu L et al (2021) Distribution characteristics of microplastics in agricultural soils from the largest vegetable production base in China. Sci Total Environ 756:143860

Zaman M, Uddin S (2024) Sewage sludge usage in agriculture: are we introducing microplastics in food chain? Soils Newslett International Atomic Energy Agency 10(2):7–8

Zeb A et al (2022) Effects of polyester microfibers (PMFs) and cadmium on lettuce (Lactuca sativa) and the rhizospheric microbial communities: a study involving physio-biochemical properties and metabolomic profiles. J Hazard Mater 424:127405

Zhang G, Liu Y (2018) The distribution of microplastics in soil aggregate fractions in southwestern China. Sci Total Environ 642:12–20

Zhang D et al (2016) The status and distribution characteristics of residual mulching film in Xinjiang, China. J Integr Agric 15(11):2639–2646

Zhang S et al (2020a) Distribution of low-density microplastics in the mollisol farmlands of northeast China. Sci Total Environ 708:135091

Zhang L et al (2020b) An overlooked entry pathway of microplastics into agricultural soils from application of sludge-based fertilizers. Environ Sci Technol 54(7):4248–4255

Zhang X et al (2021a) Systematical review of interactions between microplastics and microorganisms in the soil environment. J Hazard Mater 418:126288

Zhang Q et al (2021b) Effect of polystyrene microplastics on rice seed germination and antioxidant enzyme activity. Toxics 9(8):179

Zhang Y et al (2022) Effects of microplastics on soil carbon dioxide emissions and the microbial functional genes involved in organic carbon decomposition in agricultural soil. Sci Total Environ 806:150714

Zhou B et al (2020) Microplastics in agricultural soils on the coastal plain of Hangzhou Bay, east China: multiple sources other than plastic mulching film. J Hazard Mater 388:121814

Zhu F et al (2022) Microplastics altered soil microbiome and nitrogen cycling: the role of phthalate plasticizer. J Hazard Mater 427:127944

Chapter 2
Microplastics in Crop Systems

Nazima Habibi, Saif Uddin, and Montaha Behbehani

Abstract The pervasive presence of microplastics (MPs, <5 mm) and nanoplastics (NPs, <100 nm) in agricultural systems poses significant risks to crop productivity, food security, and human-animal health. The terrestrial environments—where plastic contamination is substantially higher—remain understudied, particularly in the context of crop systems. This chapter synthesizes current knowledge on the uptake, translocation, and phytotoxic effects of micro-nanoplastics (MNPs) in crops, emphasizing the interplay of particle characteristics (size, polymer type, surface charge, and dose), crop species, developmental stages, and environmental exposure pathways. Smaller NPs (e.g. ≤100 nm) exhibit greater bioavailability, infiltrating roots via apoplastic transport, endocytosis, or crack-entry modes, while foliar uptake occurs through stomata and trichomes. Analytical challenges, such as organic matrix interference during digestion, limitations in spectroscopic resolution (e.g. Fourier transform infrared spectroscopy (FTIR), Raman), and difficulties in distinguishing smaller MPs and NPs from natural particles, hinder accurate quantification. Advanced techniques, including pyrolysis gas chromatography-mass spectrometry (Py-GC/MS), inductively coupled plasma mass spectrometry (ICP-MS), and synchrotron-based imaging, show promise but require standardization, and the information obtained is incomparable to existing data where mostly numbers are reported instead of mass. The toxicity is size dependent, and these techniques lack the capability to provide any size information. Mechanistic studies reveal species-specific responses: dicots (e.g. lettuce, carrot) and certain legumes (e.g. cowpea) accumulate NPs more readily than monocots (e.g. wheat, maize), with polymer-dependent toxicity observed for poly(vinyl chloride) (PVC), polyethene (PE), and polystyrene (PS). Critical gaps persist in understanding real-world NP interactions with aged/weathered plastics, multi-contaminant effects, and long-term ecological impacts. The absence of harmonized protocols and field-validated risk assessments limits regulatory frameworks. This chapter underscores the urgency of

N. Habibi (✉) · S. Uddin · M. Behbehani
Environment and Life Science Research Centre, Kuwait Institute for Scientific Research, Safat, Kuwait
e-mail: nhabibi@kisr.edu.kw

S. Uddin, M. Zaman (eds.), *Micro- and Nano-plastics in Soil and Crop Systems*, Sustainability Sciences in Asia and Africa, https://doi.org/10.1007/978-981-95-2740-3_2

interdisciplinary efforts to refine detection methodologies, elucidate translocation mechanisms, and evaluate food chain contamination risks, thereby informing strategies to mitigate plastic pollution in agroecosystems.

Keywords MNPs · Organic matter · Translocation mechanism · FTIR · Raman

2.1 Introduction

Over the recent decade, research on microplastics (MPs) and nanoplastics (NPs) has grown dramatically. Nonetheless, marine organisms and the environment have been the focus of most of the research. Consequently, more marine animal species than freshwater or terrestrial species have been examined for the presence of MPs (<5 mm) and NPs (<100 nm) (Horton et al. 2017; Alimi et al. 2018; Hartmann et al. 2019). Although the annual release of MPs is 4 to 23 times higher in terrestrial environments compared to the aquatic environment (Horton et al. 2017), higher plants and crops have been very limitedly explored to monitor MP and ecotoxicological assessments (Yu et al. 2024; Jadhav and Medyńska-Juraszek 2024; Iqbal et al. 2023).

Plants form the primary source of human food. The Food and Agriculture Organization (FAO) lists 350,000 plant species, of which about 8.6% (30,000) are edible for humans, with at least 7,000 species in cultivation (Khoury et al. 2022a). In yet another report of FAO, a handful of staple crops (n = 255) including cereals (n = 26), roots and tubers (n = 17), pulses (n = 26), vegetables (n = 44), fruits (n = 69), nuts (n = 14), oils (n = 28), herbs and spices (n = 24), and sugars (n = 3) and stimulant crops (n = 4) provide the bulk of global food supply (Bélanger and Pilling 2019). Evidence reports that these small groups of globalized crops have increased dietary homogeneity; however, a decline in their genotypic and phenotypic diversity has been recorded in the farmer's field over the past century (Khoury et al. 2014, 2022b). Further reports indicate a reduction in crop production when exposed to emerging contaminants like MPs and NPs (Pérez-Reverón et al. 2022).

Microplastics were confirmed as a risk to edible crops, and their movement and bioaccumulation within the food chain are now known (Xu et al. 2022). It is essential to comprehend the movement of micro-nanoplastics (MNPs) in plants since this information sheds light on their fate and pathways within the plant system. The presence of plastics affects the ecosystem as a whole, and its movement within plants acts as a vector for other hydrophobic contaminants, facilitating their extensive dispersion. From 2015 onwards, the majority of research is on the phytotoxicity of microplastics to plants (Gong et al. 2021). Assessment of MP uptake is still nascent. In this chapter, we have summarized the key findings of these studies, the parameters affecting MP uptake, analytical methods to track the uptake, and the major mechanism of MNP absorption as well as translocation within the crops. Major gaps have also been discussed.

2.2 Factors Affecting Microplastic Uptake by Crops

Several studies have reported the adherence and absorption of MPs in crops. It has been believed that MPs cannot be directly internalized into plant tissue because they are too big to fit past the physical boundaries of intact plant tissue (Yu et al. 2024; Mateos-Cárdenas et al. 2021; Dehghanian et al. 2023). However, smaller MPs and NPs can be absorbed by plants in the same way as micronutrient particles are absorbed. Chemical and physiological barriers are among the obstacles that MPs must overcome to enter plants. MNPs form aggregates surrounding the plant tissues and can be adsorbed onto plants through various routes. An experimental study observed 50–4800 nm-sized MNP aggregates on the seeds of *Lepidium sativum* during germination (Bosker et al. 2019). These MNPs have adhered to the root hairs of seeds that passed the germination stage. In yet another study, polystyrene nanoplastics (PSNPs) predominantly adhered to the root surface, and no signs of uptake and translocation were recorded in root stocks of citrus plants grown aeroponically (Hussain et al. 2025) except for alterations in the external root cell layers. The adherence created root apoplastic barriers that inhibited nutrient uptake. The structure and composition of phyllosphere communities were altered due to PSNP adsorption on tomato leaves (Shi et al. 2023). The presence of trichomes facilitated the accumulation of smaller-size PS-MNPs on leaf surfaces of maize and soybean plants (Guo et al. 2023). The deposition altered the epicuticular wax layer from crystalline to amorphous form.

Very few studies have demonstrated the uptake of NPs by plant tissue. MP or NP uptake in plants depends on plant size, species, particle charge, particle concentration, and exposure duration.

2.2.1 Size

The particle size plays an important role in the absorption of the MNPs by the plant cells. Few studies have proven that smaller MPs and NPs are more readily absorbed by the pores between the cell walls and internalized through the vascular bundles (Yu et al. 2024). A study revealed that smaller-sized PS-NPs (50 and 100 nm) readily penetrated the root epidermis of taro (*Colocasia esculenta*), whereas PS-MPs (200 and 500 nm) were not detected in the tissues (Xu et al. 2025). Hence, it suggests that a smaller-sized plastic particle is easily uptaken. It also minimizes the water uptake as it blocks the pores, thus decreasing the germination rate. Studies on cucumber plants treated with NPs of 100, 300, 500, and 700 nm size revealed that 100 nm particles were readily engrossed by the roots and transferred to the shoots. The reason behind the easy translocation of smaller particles was a smooth passage between the cellulosic cell walls and the condensed pectin network (Li et al. 2021a).

The uptake of MNPs in maize seedling roots exposed to polystyrene beads of 0.2, 0.5, and 1.0 μm (NPs) and 2.0 and 5.0 μm (MPs) was investigated in a microcosm (Li et al. 2023a). The findings demonstrated that under particle sizes in root and shoot biomass stayed constant. Compared to MPs, the NPs significantly boosted their distribution and accumulation in roots. When treated with 5.0 μm PSMPs, none were found in any root tissues, and the PSMPs of 2.0 μm size were mostly found in the xylem vessels (Li et al. 2023a). Similar to this report, it was further demonstrated that 100 nm PSNPs accumulated in the root tissue of fava beans and reportedly blocked the cellular pores, potentially obstructing nutrient transport (Jiang et al. 2019). Particles 5 μm in size were not observed in the root tissue.

In yet another investigation, carrot roots were studied for PSNP absorption and translocation (Dong et al. 2021). The group revealed that 1 μm MNPs entered the roots and accumulated in the intercellular layer. However, these could not enter the cells. Those sized 0.2 μm migrated up to the leaves, while the particles >1 μm entered the roots only in the presence of arsenic. This suggests the possible role of other environmental contaminants along with size in the absorption and uptake of MNPs. The study further reported arsenic increased the negatively charged area of PS and also distorted the cell walls, allowing the smooth uptake of PS particles <200 nm.

Yet in another study, strawberry seedlings were exposed to 100 and 200 nm PSNPs where both sized particles were detected in the vascular bundles of the petioles after 7 days of exposure. On the 14th day, 100 nm NPs passed the petioles, whereas 200 nm could not be directly observed in the strawberry seedlings, strongly suggesting that MP absorption and translocation also depend on the size of vascular bundles (Zhang et al. 2023). It was also reported that different sizes of MPs not only influence their internalization but also the physiological responses and molecular mechanisms in plant species. PSNPs of 100 and 500 nm size, when exposed to lettuce roots, were absorbed and translocated to stems and leaves with a stronger movement ability of smaller NPs. The physiological, biochemical, and genetic parameters were affected to a greater extent by larger NPs (Hua et al. 2024).

A general observation from the above studies is that while NPs can enter plant cells directly, larger MPs find it difficult to get through the tissues of vascular plants due to the plant's cellular structure. Also, the thickness of the cell wall and epidermis varies according to crop type, environmental parameters, contaminants, growth stage, etc., which define the translocation of NPs and MPs across these barriers. Consequently, the size of MNPs that will be absorbed by the plants will be different under different experimental conditions.

2.2.2 Species

Prior studies have indicated that distinct plant species may employ distinct coping mechanisms to manage various environmental stressors (Gong et al. 2021). The uptake of nutrients and defensive action against the emerging contaminants are diverse for diverse species. The responses of several plant species to identical MNP exposure are not well understood. Therefore, to make intelligible conclusions on the impact of MNPs on plants, a thorough analysis of the reactions of different plant species under the same experimental conditions is essential. An investigation on seeds of four plants (Italian lettuce, radish, wheat, and corn) exposed to PSNP and PSMP suspension reported that the tolerance and sensitivity towards the MPs differed according to the crop type (Gong et al. 2021). The Italian lettuce plant showed maximum sensitivity towards polystyrene exposure. The root/shoot ratio, dry weight, and length of Italian lettuce and corn were dramatically reduced, whereas the radish and wheat plants remained unaffected. This can be attributed to the difference in root exudates produced by the monocots (wheat and corn) and the dicots (Italian lettuce and radish). Similar responses were recorded by Zantis et al. (2023), in which lettuce and carrot plants were significantly affected, whereas barley and wheat remained unaffected (Zantis et al. 2023). A significant delay in seed germination of lettuce was recorded, whereas the biomass of carrots was decreased.

The absorption and transportation of MNPs encounter numerous physical and chemical obstacles in plants that vary according to crop species and growth stages. The structure and obstacles vary according to the internal structure of plants. A comparison between wheat and maize revealed the latter absorbs more MNPs than the former (Erdem et al. 2023). The deposition was in different tissue types as well. In maize, larger PP particles were recorded in the stem, whereas smaller PPs were absorbed by wheat roots. In another study, it was observed that the MNPs can be absorbed during the seed germination stage; however, the early development of monocot species was not affected by PSMP exposure (Zantis et al. 2023). Rice and peanuts absorbed PSNPs at the seed-setting stage and accumulated them at the grain maturation stage (Jiang et al. 2022). The MPs entered pea plants during the root development stage while Casparian strips were forming (Kim et al. 2022).

However, contrary to the above, irrespective of the species, large MPs were found associated with leaf surfaces and trichomes both in maize and in soybean plants (Guo et al. 2023).

In addition to this, below-ground crops such as tubers, bulbs, and leguminous crops such as peanuts are more vulnerable to MNPs present in the soil. Peanuts absorbed PSNPs through their roots and reached the grains at the maturation stage. PSNPs at a concentration of 250 mg kg^{-1} reduced the average seed weight of peanuts and also reduced the overall nutritional quality, minerals, amino acids, and unsaturated fatty acid content of the seeds (Jiang et al. 2022).

Different polymers are absorbed differentially by plant species. A total of 11 different plastics, including polystyrene (PS), polyethene (PE), polypropylene (PP), polymethyl methacrylate (PMMA), poly(vinyl chloride) (PVC), poly(ethylene terephthalate) (PET), polycarbonate (PC), polyamide 6 (PA6), polyadiohexylenediamine (PA66), polylactic acid (PLA), and poly(butyleneadipate-co-terephthalate) (PBAT) were exposed to cowpea, flowering cabbage, rutabagas, and chieh-qua. PVC, PET, PE, and PA66 could be found in all crop samples. The predominant components in the samples of all four crop species were PVC and PE, although cowpea samples had significant concentrations of PET. Additionally, the accumulation levels of different edible crop parts varied significantly; for example, in blooming cabbage samples, stems accumulated more NPs than leaves, while in cowpea samples, peas were higher in NP content than pods (Ye et al. 2024).

It becomes quite evident that MNP uptake and translocation are species dependent. The crop itself grows in a complex environment, and each species is unique in structure and function. Experiments with multiple species and an array of polymers in field-realistic conditions would shed more light on the transmission of MNPs in terrestrial ecosystems.

2.2.3 Particle Dose

The effects of MNPs on plant systems are intricate and modulated by several factors, including exposure dose (Shi et al. 2024). A meta-analysis of 37 studies on MP uptake concluded that higher doses of smaller MPs exert more pronounced inhibitory effects on root activity and biomass (Shi et al. 2024). The effect of low (10^3), medium (10^5), and high (10^7) particles per ml was investigated on four crop types (Zantis et al. 2023). The highest concentration delayed the germination rates of lettuce, wheat, and carrot. The root length, shoot length, and biomass of germinated wheat seedlings were also reduced. Similarly, shoot length and bud biomass of carrot seedlings also decreased at the highest concentration of PSNPs. Further growth of the crops was monitored, and a reduction in the wet weight of wheat was observed (Zantis et al. 2023). Inhibition of root growth was recorded in onion seedlings exposed to 0.1 and 1 g L^{-1} of PS-NPs as compared to a 0.01 g L^{-1} concentration (Giorgetti et al. 2020). Rice and peanut grains absorbed PSNPs at a concentration of 250 mg kg^{-1} (Jiang et al. 2022). Overall seed-setting rate and seed weight were significantly reduced when rice and peanut crops absorbed NPs at the grain maturation stage.

Higher concentrations of MP considerably decrease plant growth and physiological activities, indicating a positive correlation between exposure dose and toxicity. Some researchers have shown that MPs can affect physiological and biochemical traits as well as plant growth parameters through a variety of mechanisms. For example, a lower concentration of MPs ($\leq$5 mg·L^{-1}) increased root hydraulic conductivity, and higher PS treatment (200 mg·L^{-1}) reduced the chlorophyll a, b, and total chlorophyll content in wheat seedlings (Li et al. 2023b). Three concentrations (100, 300, and 500 mg L^{-1}) of carboxyl-tagged PSNPs invaded cotton roots, causing upregulation of genes related to antioxidant activity and inducing oxidative stress. Significant upregulation of genes involved in antioxidation was observed at a higher dose of 300 mg/L PSNPs in the above experiment (Li et al. 2024). The PS-COOH treatment triggered the phenylpropanoid biosynthesis pathway, in turn stimulating the build-up of flavonoids and lignin. An altered sucrose metabolism was also observed in response to PS-COOH exposure at a higher concentration (Li et al. 2024). Within the plant, translocation is also dose dependent, as studied by Li et al.

(2019) who found that there were only a few sporadic fluorescence signals in the tissues of lettuce stems and leaves, with PSNPs sticking to one another in the vascular bundle, exhibiting 'grape' or 'chain' aggregation, and microbeads adhering to one another in the leaf tissue in the form of dispersion. This implies that the vascular bundle can only carry trace amounts of MNPs to the vascular systems of stems and leaves (Li et al. 2019).

MPs are ubiquitous not only in the terrestrial environment but also in the atmosphere (Habibi et al. 2022). Atmospheric MPs have been absorbed by the plant leaves through the stomatal and cuticular openings (Shi et al. 2023; Guo et al. 2023; Lian et al. 2021; Sun et al. 2021). In a recent study, a dose-dependent response was monitored by Li et al. (2025). They reported the widespread occurrence of PET and PS polymers as a ruling factor in their absorption by leafy plants under field conditions. The concentrations were higher in the leaves of plants grown in locations near the Dacron factory proximal to a landfill site as compared to those grown in parks and university campuses (Li et al. 2025). In addition to field plants, they experimented on leafy vegetables (*Brassica rapa* var. *glabra*, *L. sativa* var. *ramosa*, and *B. rapa* var. *chinensis*) planted in open-air fields and under greenhouse conditions. Likewise, the field-grown plants and the vegetable crops grown in open fields accumulated more PET and PS than the greenhouse vegetables. It was also observed that the outer leaves of these vegetables absorbed higher quantities of PET and PS than the inner leaves. Li et al. (2025) also demonstrated the maize leaves exposed to dust collected from the Dacron factory (containing 1.05×10^3 μg g^{-1} of PET) in a microcosm experiment accumulated five and seven times higher magnitudes of PET at the 5th and 10th day, respectively. Upon halving the dust concentration, almost double the amount of PET was absorbed on the 10th day. These observations indicate that plant leaves are the primary unit of vegetable crops, and the likelihood of MPs entering the food chain is greater via this pathway. This also warrants investigations on MP accumulation in herbivores and humans, taking into consideration the health and ecological risks.

In experimental conditions, it is difficult to mimic the environmental concentrations of MNPs. Nevertheless, besides a range of particles available at varied densities, the plants in a terrestrial environment are under constant chronic exposure, and this should be taken into consideration while estimating the risk to the plants and indirectly to their consumers.

2.2.4 Surface Chemistry/Charge

Environmental elements like ultraviolet radiation, heat radiation, chemical oxidation, and biodegradation cause the MNPs to undergo a variety of surface changes. Carboxyl groups (-COOH) are the most prevalent functional groups formed due to surface oxidation and functionality acquisition. Additionally, plants are more likely to absorb carboxyl-modified polystyrene microplastics (PS-COOH) with a negatively charged surface. It thus becomes important to investigate the uptake of MPs with carboxyl-tagged groups or charged particles by plants. Tomato leaves were exposed to differently charged nanoplastics, where positively charged NPs entered the leaf tissue and dispersed to the mesophyll cells with more devastating effects (Shi et al. 2023). They caused growth inhibition, an increase in antioxidant enzyme activity, and altered gene expression. PSNPs tagged with amino and carboxyl groups were absorbed by rice seedlings. Post-absorption, PS-COOH affected ion transport and terpene biosynthesis in rice seedlings, whereas PS-NH_2 impacted the synthesis of macromolecules (Wang et al. 2022). Carboxyl-modified PSNPs were found invading cotton roots (Li et al. 2024). The treatment with PS-COOH activated the phenylpropanoid biosynthesis pathway, resulting in flavonoid and lignin accumulation, as well as modified sucrose catabolism (Li et al. 2024).

Arabidopsis thaliana collected both positively and negatively charged NPs. The uptake of amino-modified polystyrene nanoplastics with positive surface charges was restricted as the media and root exudates facilitated the aggregation of particles. Therefore, compared to negatively charged sulfonic-acid-modified nanoplastics, positively charged nanoplastics accumulated at relatively low levels in the root tips, but they also caused a greater accumulation of reactive oxygen species and more effectively impeded plant growth and seedling development. In contrast, the apoplast and xylem were frequently found to contain negatively charged NPs. These findings provided direct evidence that NPs can accumulate in plants, depending on their surface charge (Sun et al. 2020).

A quantitative estimation of MP bioaccumulation in lettuce leaves indicated the toxicity was largely determined due to the surface functional group and not only the accumulating co-particles (Xu et al. 2022). Contrastingly, the trace element copper as an additive to the soil, along with PSNPs, caused a reduction in its uptake at the 7th and 42nd day of exposure in pea plants (Kim et al. 2022).

2.2.5 Exposure Time

The absorption and translocation of MPs differed according to the duration of the exposure. Longer exposure times typically result in higher MP accumulation and may be more pronounced detrimental effects on plant growth, even if the effects of MP exposure on plants are complicated and dependent on a number of variables. MNPs were absorbed by the plants after seven days of sowing, indicating the ability of crops to uptake them at a very early growth stage (Gong et al. 2021). Also, PSNPs of 100 and 200 nm size were observed till the petiole of the strawberry seedling on the 7th day. On the 14th day, direct observation of PS-NPs was not evident (Zhang et al. 2023). Pea crops grown in MP-contaminated soil for 7 days showed 110% absorption of MPs (Kim et al. 2022). A further chronic exposure maintained 107% absorption. PS beads were found in the stele on the secondary roots after 12 hours and in the wheat's vascular tissue and epidermis after 48 hours (Li et al. 2020). A dose-dependent MP uptake and effect was also recorded in cotton roots (Li et al.

2024). A 7-day exposure of PS-NPs to the roots of rice seedlings favoured the uptake of particles (Li et al. 2023a), while a 10-day exposure facilitated the accumulation of smaller PSNPs.

2.2.6 Polymer Type

The uptake of microplastics (MPs) and their subsequent effects on plants are greatly influenced by the type of polymer they absorb. Polymer type is expected to affect the physiological and metabolic systems of crops. Most of the studies have explored the effects of PS particles. These have proven adverse effects on plant growth, nutrient uptake, photosynthetic capability, and other physiological parameters. Other polymer types have frugally been investigated. Among different plastics, such as PS, PE, PP, PMMA, PVC, PET, PC, PA6, PA66, PLA, and PBAT, only PVC and PE were the dominant NPs accumulated in four crop types. Compared to stem and melon crops, the degree of NP enrichment in legume and leaf crops was higher (Ye et al. 2024). Shi et al. (2022) reported the absorption of PE spheres in cortical tissues of roots and stems of sweet potatoes. The prevalence of PET or PS particles accumulated in field-grown plants, leafy vegetable crops, and maize leaves was reported (Li et al. 2025).

The effect of real-world MNPs polyethersulfone (PES), polyamide (PA), polypropylene (PP), low density polyethylene (LDPE), polyethylene terephthalate (PET), polyurethane (PU), polystyrene (PS), polycarbonate (PC) of different shapes (fibres, films, foams, and fragments) was investigated on the growth of carrot seedlings (Lozano et al. 2021). Both the shape and type of polymer affected the shoot biomass. It increased when exposed to all the polymer types except PS and PA. PET films recorded the highest biomass, followed by LDPE foams and PP films. Fragments, in general, also increased the shoot biomass. Foams, particularly LDPE and PS, reduced the shoot biomass. Similar effects were observed for roots, with double the intensity. PU foam exerted the maximum positive effect on root biomass, followed by LDPE films, PP fragments, and PET films. Irrespective of the shape, the MPs increased both the root and shoot biomass in carrots. Real-world MNPs comprise not only different types but also additives such as plasticizers, blowing agents, and stabilizers to provide flexibility, roughness, durability, etc., which might also play a role in perturbing the plant growth.

2.3 Analytical Methods for Visualizing and Tracking MP and NP Uptake

Comprehending the uptake and translocation of MNPs by plants has grown in importance. Accurate tracking and quantification of MNPs in plants is made possible by the development of microplastic detection methods, including optical and electron microscopy, in addition to spectroscopy. Each of these methods has special benefits for illuminating the biophysical and biochemical relationships between plant tissues and MNPs (Hua et al. 2023). The selection of detection techniques is critical to the precise identification of MNP absorption and translocation. The diverse size, shape, and polymer types make it difficult to standardize detection techniques. For the effective extraction of MNPs from plant tissues, researchers are looking into a number of approaches, including chemical and enzymatic digestions and density separation to segregate MNPs from the matrix.

Observing the absorption and transport process of microplastics in plants under a microscope presents numerous challenges as well. For example, the small size and variety of shapes of MNPs make it challenging to identify and differentiate them from plant matrices. MNPs have been found in tissues using conventional microscopic methods like light and electron microscopy; however, these methods can only provide morphological information, as both natural fibres and polymers closely resemble one another morphologically and often lead to false positives.

2.3.1 Factors Affecting the Analytical Ability

2.3.1.1 Plant Complexity

The extraction of MNPs from the plant matrix is a crucial step. The digestion of organic matter in plant tissues for MNP extraction presents significant technical challenges due to the complex matrix of plant materials, which includes cellulose, lignin, and other recalcitrant biopolymers. Conventional chemical digestion methods, such as alkaline (e.g. KOH), acidic (e.g. HNO_3), or oxidative (e.g. H_2O_2) treatments, often fail to fully degrade the plant's organic components without compromising the integrity of certain plastic polymers. Hurley et al. (2018) in their study highlighted that incomplete digestion can leave residual organic particles, complicating MNP identification via spectroscopic techniques. Additionally, aggressive reagents like HNO_3 or high-temperature treatments risk degrading susceptible polymers such as PET, PA, (Cole et al. 2014) or polylactic acid (PLA) leading to underestimation of MNP in organic-rich matrices. Enzymatic digestion, while gentler, struggles with efficiency in lignocellulosic-rich tissues, requiring prolonged incubation times and is likely to leave plant residues (Loder et al. 2017). Moreover, validation of digestion protocols is exacerbated by the variability in plant tissues, necessitating species-specific optimization to avoid misidentification. These issues underscore the need for standardized, non-destructive methods tailored to plant matrices to ensure accurate MNP quantification.

2.3.1.2 Miscellaneous Transport Mechanisms

An additional degree of complexity is introduced by the different transport pathways found in plant cells and tissues. Because MNPs can transfer across various parts of a plant, i.e. between the roots, stem, leaves, and fruits, it is challenging to elucidate the distribution and movement of MNPs in the plant system. Nevertheless, it is still unclear what the most dominant mechanisms governing the uptake, transport, and storage of MNPs in plant tissues are. It becomes imperative to conduct studies to identify MNPs in different plant parts to have a comprehensive understanding of the scale of MNP pollution in plants and to underpin regulatory mechanisms for MNP reduction in food chains.

2.3.1.3 Lower Concentration Within Plant Tissue

Another obstacle to accurately quantifying and identifying MNPs in plant tissues is the lower occurrence. Existing detection techniques such as microscopy and spectroscopy, differ in terms of resolution, sensitivity, and specificity. At times, they fail to provide high confidence estimations whether, the MNPs reported are a result of underestimation or the uptake mechanism itself. Consequently, it becomes crucial to distinguish MNPs from background noise and to develop techniques that can achieve lower detection limits. Some of the techniques used so far include micro-Raman spectroscopy, micro-Fourier transform infrared spectroscopy (FTIR), Attenuated Total Reflectance Fourier-Transform Infrared spectroscopy (ATR-FTIR), laser direct infrared (LDIR) imaging, Thermogravimetric Analysis (TGA), and optical and scanning electron microscopy, to name a few.

2.3.2 Microscopic Methods

2.3.2.1 Optical Microscopy

Although simple, cost-effective, and ideal for preliminary observation, its restricted resolution would not make it suitable for identifying extremely tiny particles (González-Melendi et al. 2008). Besides, the MP identification with optical microscopy is based on morphological characteristics (Uddin et al. 2022) which cannot provide polymeric confirmation, and at best one can say possible MPs.

2.3.2.2 Scanning Electron Microscopy

Scanning electron microscopy (SEM) can produce high-resolution images and decipher morphological details of MNPs and plant tissues. In a study, SEM was coupled with fluorescent imaging to trace the uptake and translocation of PSNPs in rice and peanut grains (Jiang et al. 2022). Another study demonstrated tracking of 200 nm PSNPs in wheat seedlings using hyperspectral-enhanced dark-field microscopy coupled with SEM (Li et al. 2023b). PSNP absorption and translocation from leaves to roots were traced using SEM and transmission electron microscope (TEM) in lettuce leaves (Lian et al. 2021). PS beads were located on the stele of the secondary roots after 12 hours, and in the wheat's vascular tissue and epidermis after 48 hours, by SEM (Li et al. 2020) and PS-NPs were confirmed via SEM in dried cucumber tissue (Li et al. 2021b). The absorption of PS-NPs tagged with Europium lanthanides was observed in tomato leaves, and their transportation to roots was traced using SEM (Shi et al. 2023).

2.3.2.3 Confocal Laser Scanning Microscopy

It is a powerful technique for tracking the transport pathways of MNPs and visualizing their spatial localization within plant tissues. This approach commonly uses fluorescent dyes to mark the particles and creates a three-dimensional image of the MNP embedded within the plant tissue. Fluorescence signals assist in tracking the MNPs absorbed in plant tissues. Confocal microscopy, in combination with a fluorescent dye (Nile red), was used to demonstrate the uptake of NPs by the root tissue of cucumber and its further transportation to stem, leaves, and fruits (Li et al. 2021a). Absorption of PS-NPs in rice roots was also confirmed by confocal microscopy (Wang et al. 2022). Fluorescent signals of PS-NPs were observed in BY-2 cells under a confocal microscope (Bandmann et al. 2012).

2.3.2.4 Transmission Electron Microscopy (TEM)

TEM has limited application in the detection of MPs or NPs as extensive sample preparation steps are involved, as well as the restricted field of vision. Yet the morphology and primary size of PS-COOH were visualized via TEM imaging in cotton crops (Li et al. 2024). To assess the actual size of the PS microspheres, a drop of the nano-PS solution (1 g L^{-1}) was applied on TEM grids coated with formvar, allowed to settle, and allowed to dry. The major and minor axes of TEM images were measured using the ImageJ tool to assess the particle diameters (Giorgetti et al. 2020). TEM was used to monitor the ultrastructural changes caused by the absorbed NPs in tomato leaves (Shi et al. 2023).

2.3.3 Spectroscopy

Spectroscopic techniques, mainly vibrational spectroscopy (Raman and infrared), are commonly used techniques for microplastic analysis in environmental samples.

2.3.3.1 Fourier-Transform Infrared Spectroscopy

Fourier-transform infrared spectroscopy (FTIR) is often used to identify the polymer type within environmental samples, including biological samples that have been digested. Although the method is mostly used to find the build-up of MNPs in marine biota, it is still not widely used for evaluating plant samples.

2.3.3.2 Raman Spectroscopy

Raman spectroscopy is another approach to molecular characterization. MPs can be distinguished from other particulate matter using Raman spectroscopy. Similar to FTIR, this method provides accurate identification, but it is particularly useful when combined with microscopic techniques for a more thorough investigation. Micro-Raman spectroscopy is non-destructive in most cases and known to be effective for characterizing particles up to 1 μm size.

2.3.3.3 Advanced Coherent Anti-Stokes Raman Scattering

More recently, the highly advanced coherent anti-Stokes Raman scattering is being observed as a novel tool to track the uptake of NPs by plant cells (Liese et al. 2024).

2.3.3.4 Inductively Coupled Plasma Mass Spectrometry (ICP-MS)

This method provides information about the composition of MNPs. It provides additional information on the constituents present in the MNPs as well as its quantity in terms of mass. Luo et al. (2022) used this technology in a groundbreaking study to investigate how plants absorb MNPs. The study clarified the possible use of inductively coupled plasma mass spectrometry (ICP-MS) in environmental impact studies by successfully demonstrating its value in tracking and measuring NP absorption in plant species. Doping the NPs with lanthanide chelates offered a flexible approach for tracing the movement of MNPs within the plants (Luo et al. 2022). Using time-gated luminescence and the time-resolved fluorescence of the Eu chelate, Europium tagged Polystyrene (PS-Eu) particles in the roots and shoots of wheat and lettuce were visualized. SEM was used to confirm the presence of PS-Eu within the plant roots. The researchers examined the uptake of these PS-Eu particles by wheat and lettuce using quantitative analysis with ICP-MS. Further, this technique was used to confirm the measurable levels of NPs absorbed by cucumber plants (Li et al. 2021b). The NPs were tagged with palladium.

2.3.3.5 Pyro-GC Coupled with Mass Spectrometry (Py-GC/MS)

Pyrolysis gas chromatography-mass spectrometry (Py-GC/MS) shows promise as a method for measuring NPs, especially in samples with complex environmental matrices and low concentrations (Yu et al. 2024). This approach provides significant benefits for the precise detection and measurement of all the plastics, including nanoplastics. Using cucumber (*Cucumis sativus*) as the model species, a novel method was devised to measure the uptake of nanoplastics in plants by combining alkaline digestion, cellulose precipitation, and ultrasonic leaching, followed by Py-GC/MS analysis (Li et al. 2021b). The chemical composition and mass concentration of NPs in various crops (cowpea, flowering cabbage, rutabagas, and chieh-qua) were ascertained using a method that combined nitric acid digestion, multiple organic extractions, and pyrolysis gas chromatography-mass spectrometry (Py-GC/MS) quantification.

The chemical composition and mass concentration of NPs in various crops (cowpea, flowering cabbage, rutabagas, and chieh-qua) were ascertained using a method that combined nitric acid digestion, multiple organic extraction, and Py-GC/MS quantification. Although the accumulation levels varied considerably due to the crop type, the results demonstrated that PVC, PET, PE, and polyadiohexylenediamine (PA66) NPs could be absorbed by the crop species (Ye et al. 2024).

Contrary to the above two studies, due to the presence of naturally occurring styrene monomers in plant roots (wheat and Arabidopsis) and inadequate detection limits, Py-GC/MS was not successful. In each species' root surface, 40 nm and 1 μm polystyrene spheres were accumulated, especially at the root tip. Nevertheless, the interior root structure showed no signs of plastic particles (Taylor et al. 2020).

2.3.3.6 Micro-X-Ray Fluorescence (μ-XRF)

An innovative analytical method for examining the interactions between plants and manmade nanomaterials is μ-XRF. By offering lateral resolution in multi-elemental detection, μ-XRF provides crucial information about the fundamental processes of these elements' absorption, movement, and accumulation in plant systems. Additionally, X-ray computed tomography is a non-destructive imaging technique that makes it possible to rebuild the scanned specimens in three dimensions. The study of plant root properties has made considerable use of μ-XRF due to its ability to efficiently negate surface artefacts and need minimal sample preparation. To see the NPs in wheat roots, a synchrotron (μXRF) was used. The uptake and translocation of NPs to the shoots could be quantified. The NPs were collected in the root tip and the root maturation zones along the epidermis. Globular formations were detected by synchrotron-computed nanotomography (nanoCT), but as they were seen in both the treated and control roots, they could not be recognized as nanoplastics. An inorganic tracer (Pd) was doped with the NPs that helped elucidate the plant NP interactions (Del Real et al. 2022).

2.4 Mechanism of MP-NP Uptake in Crops

Since plant tissue is essentially a complex mixture of biopolymers, and several parameters influence MNP uptake and transport in plants, it is difficult to understand the precise mechanisms and ascertain their fate within a species experimentally. Despite the lack of research on this aspect, four mechanisms, i.e. apoplastic transport, crack-entry mode, endocytosis, and stomatal entry, are believed to play roles in the uptake and translocation of MNPs in crops (Yu et al. 2024; Azeem et al. 2021). A brief overview of each mode is given in the following section. The internalization of NPs into plant roots and their subsequent migration to the aerial portion are deemed possible (Hua et al. 2023).

2.4.1 *Endocytosis*

It has been demonstrated that endocytosis is a successful mechanism for MNPs to enter cells. Phagocytosis, pinocytosis, and receptor-mediated endocytosis are the three categories of endocytosis that may be distinguished based on the size of the material entering cells and the endocytosis process. Cells swallow solid particles, including PSNPs, into vesicles known as phagosomes through a process known as phagocytosis. A study established that PSNPs can be taken up by cultivated sycamore cells and delivered to the core vacuoles through phagocytosis.

Phagocytosis is more commonly employed by immune cells to absorb and eliminate invading particles in humans; however, this has also been seen in plant cells (Etxeberria et al. 2006).

The vacuoles of sycamore cells cultured with fluorescence-labelled MNPs for 12 hours displayed numerous bright spots of fluorescence signals, as observed under a confocal microscope. This suggested that the vacuoles were the final destination of MNPs following endocytosis (Etxeberria et al. 2006). It was also observed that PS-NPs cluttered strongly and persistently adhered to the cell walls. This significantly hindered PS-NP absorption. This challenge was overcome by combining the protoplasts of cultivated sycamores with a growth medium that contained PS-NPs. Following an overnight incubation period, PS-NPs were discovered outside of several cells and in the vacuoles. However, the endocytosis of plant cells to MNPs cannot be strongly supported by this experiment as the sycamore cells lacked the cell wall as a barrier.

Numerous biological functions in animal cells, including polarity, motility, signalling control, and proper cell development, have been linked to clathrin-independent endocytosis (Sandvig et al. 2011). Building on this hypothesis, Bandmann and group had earlier demonstrated the internalization of glucose via clathrin-independent endocytosis into the protoplast of BY-2 cells (Bandmann and Homann 2012). In a subsequent study, Bandmann explored the engulfment of fluorescent PSNPs (20 nm) in BY-2 cells. Fifteen minutes following the addition of fluorescence-labelled nanoplastics, BY-2 cells showed clear brilliant fluorescence signal patches under a confocal microscope. The very mobile brilliant spots were dispersed widely throughout the cytoplasm after travelling along the cytosolic threads. The fluorescence signal distribution in the cells was not substantially different from that in the cells incubated with 20 nm nanoplastics, and the 40 nm nanoplastics were equally quickly internalized (Bandmann et al. 2012). The findings also demonstrated that BY-2 cells with cell walls may endocytose 40 nm nanoplastics; however, the size of endocytic vesicles considerably inhibits the endocytosis of some large MNPs (more than 100 nm). According to other studies, if the restrictions on the cell wall are eliminated, cell protoplasts can even internalize MNPs up to 1000 nm in diameter (Bandmann et al. 2012; Bandmann and Homann 2012).

The two experiments mentioned above demonstrated that endocytosis is a mechanism by which nanoplastics can enter plant cells. However, in both studies, the endocytosis method depended on the creation of separate cell cultures, and there is not enough data to confirm this uptake mechanism for the entire plant.

2.4.2 Apoplastic Transport

The cells in the apical meristem of plant roots serve as the first line of defence against MNPs as they regulate cell division and differentiation, have high cell activity, are highly sensitive to external stimuli, and can secrete large amounts of mucus and secretions in response to an external signal. The mucus layer essentially captures the nanoparticles, creating a barrier for internal transfer (Huang et al. 2014). Besides this, the water channel plays a key role and absorbs the particles through gaps in the cell walls of the epidermal layer, which are further transferred to the vascular bundles (xylem) (Sun et al. 2020). The external force required is supplied by transpiration, which is the process by which water is expelled from the plant through stomata on the leaf surface. Smaller particulate matter, including MNPs, is drawn with water from the roots upwards towards the shoots by the negative pressure or surface tension created in the plant as a result of water loss from the leaves. This process facilitates the transfer of MNPs with water through the porous plant cell walls. Transpiration pull plays a major role in the plant uptake and apoplastic translocation of plastic particles (Yu et al. 2024). This is the predicted pathway of NP translocation from roots to aerial parts of cucumber (Li et al. 2021a).

Apoplastic transport of fluorescently labelled PSNPs from lettuce roots to leaves was established by Li et al. (2019). It was revealed in their experiment that lettuce was unable to absorb 1.0 μm microbeads. However, a significant number of fluorescence signals were discovered within the cell walls when exposed to submicron microbeads (0.2 μm). According to the findings, the MPs must first reach the submicron level to pass through the apoplastic barrier, subsequently moving along the free spaces of root cells and entering the root cortex or even the vascular bundles. The endodermic Kasparian ring, on the other hand, prevents the apoplastic transfer from the cortex to the vascular bundle. Therefore, external forces were believed to push the MNPs via an apoplastic route to penetrate the endoplasmic plasmalemma (Li et al. 2019). In a subsequent study, the PSNPs first entered the root system and thereafter transferred to the above-ground parts in lettuce (*Lactuca sativa*) and wheat (*Triticum aestivum*) (Li et al. 2020). The PS beads were observed inside the apical meristem of the roots as well as along the entire lateral root cap. Li et al. propose that these particles in the mucilage of the root cap allow diffusion into the apical meristem tissue and enhance their perception of the cell wall, which is extremely unstable because of ongoing cell division (Li et al. 2020). Sun et al. (2020) demonstrated that MNPs can be transported through the apoplast in *Arabidopsis thaliana*. The study also showed that the apoplastic transport of MNPs is significantly influenced by their surface charge. In particular, it was found that positively charged nanoplastics (71 nm) did not behave in the same way as negatively charged nanoplastics (55 nm), which were able to effectively enter plant vascular bundles via the apoplastic pathway.

MPs infiltrated through the roots of pea plants growing in MP-exposed soil. These were further translocated to the aerial parts (Kim et al. 2022). A quantitative study on PSNP exposure from roots to shoots reported a very limited translocation (<3%) in wheat and lettuce leaves (Luo et al. 2022). It was further recorded that once absorbed via the roots and translocated to vascular bundles (xylem tissue of roots), they easily move upwards to stems and leaves with the water and nutrient flow.

Electrostatic interactions may cause positively charged plastic nanoparticles to be trapped in plant secretions, or they may adhere to regions like root hairs and edge cells that have protective negatively charged mucosal coats. Furthermore, the apoplastic transport mechanism may potentially be impacted by the chemical characteristics of MNPs. For instance, because of their poor water solubility and high octanol-water partition coefficient values, certain polymeric polymers with long alkyl chains and large molecular weights might be more challenging to absorb through the apoplast pathway. Therefore, to improve our comprehension of the extracellular transport mechanism of PS-NPs, more research on the apoplastic transport process of MNPs with various physical and chemical properties is required in the future.

2.4.3 Crack-Entry Mode

A study on wheat and lettuce plants found that PS and PMMA particles could enter the plant roots through the cracks formed during lateral root formation (Li et al. 2020). Effective absorption depends upon the polymer type and crack penetration. MNPs typically cannot diffuse into the intercellular space and pass through plant cell walls if their particles are bigger than the cell wall pores (3.5–5.0 nm) and intercellular plasmodesmata (50–60 nm). However, the ongoing cell division leads to the formation of porous tissue at the apical meristem, allowing the diffusion of MNPs. At the cell separation stage, pores are formed between the lateral root and epidermal cells, allowing the passage of MNPs into the root stele. Strong fluorescence signals were found in the cracks at lateral roots 50–100 mm from the root tip by Li et al. (2020), indicating that these ruptures serve as the main entry points for MP-NPs into the xylem of lettuce and wheat roots. Within two hours, PS and PMMA entered the xylem duct, travelled through the xylem by the transpiration pull, and finally arrived at the wheat cortical tissue twelve hours later. The vascular system of the roots and stems was the primary location for MNPs that entered through these fissures, according to confocal pictures of wheat cross sections.

It is important to remember that granular plastics' mechanical flexibility might be the key to successful absorption. Nevertheless, studies on the process of crack penetration are lacking, and the findings of Li et al. (2020) build our knowledge of a novel mode of MNP uptake and translocation. Future studies should also account for polymer type and crop species.

2.4.4 Stomatal Entry

All three of the above-described absorption and translocation mechanisms involve MNPs that are absorbed in the roots of plants and move upwards from the roots to the shoots. MPs are absorbed by the crops through the aerial parts as well. A study recently found that maize leaves can absorb NPs and transport them downwards from leaves to roots (Sun et al. 2021). Potential routes for nanoparticles to penetrate plant leaves include stomatal and cutaneous uptake. According to SEM data, PSNPs mostly showed up in and around the stomata on day 7, indicating that stomatal opening may be one way PSNPs penetrate the vasculature. Additionally, the investigation discovered that, following two days of exposure, NPs were found to be more prevalent in the stems, but less prevalent in the roots. This suggests that the sequence of NP translocation was from the leaves to the stems and then from the stems to the roots. The high concentration of PSNPs in and around the stem's vascular system supported the leaf-to-root translocation through the vascular bundle, as seen in the fluorescent microscopy images. Unlike the xylem in the roots, the phloem, the main channel for transporting nutrients and water, is responsible for PSNP migration in the case of foliar uptake. Similar observations for PSNP absorption and migration were recorded in maize leaves (Li et al. 2025). The absorbed MPs migrated from the bundle sheath to the phloem with the osmotic exchange in vascular tissue, according to high spectral imaging (his) and confocal images of transverse sections of vascular tissue of maize stem.

The possible source of MNPs for foliar stomatal uptake is aerosols. Lian and co-workers reported the absorption of PSNPs through the leaves of lettuce (*Lactuca sativa*) and its transport downwards to the roots. This reduced the micronutrients and essential amino acid content of the lettuce plant. Vascular bundles are the pathways for translocating the NPs to plant roots when entered through the leaves. This was also observed in the maize crop (*Zea mays*) (Sun et al. 2021). Foliar absorption and translocation to roots were recorded in tomato plants by Shi and co-workers

(Shi et al. 2023). In another study, foliar uptake and spread via stomatal openings were recorded in maize and soybean seedlings (Guo et al. 2023). However, the foliar uptake is largely defined by the leaf characteristics (Li et al. 2025). High levels of PET polymer were found in the leaves of *Hibiscus syriacus* (1.23×10^5 ng per g DW) and *Brassica papyrifera* (5.32×10^4 ng per g DW) out of the 12 plants that were examined at the Dacron factories. This was thought to be connected to faster respiration in these plants, which caused a greater flux of airborne MNPs to enter the leaves. Particles are more likely to enter leaves directly by gas exchange when transpiration rate and stomatal conductance are high, showing that leaves have a high stomatal exchange capacity. This implies that one of the primary pathways for absorbing PET is stomatal respiration. In addition, the size of stomatal pores in different plant species is an order of magnitude higher than the atmospheric MNPs, which allows an easy passage into the leaves. This was well demonstrated in the stomatal closure experiment of Li and group (Li et al. 2025).

To demonstrate the passage of MNPs from leaves to other plant parts, a simulation experiment was conducted on maize seedlings. The Eu-labelled PSNPs first accumulate on the leaf surface and then are internalized via the stomatal openings, migrating to the mesophyll cells and further into the trichomes. The absorption was quantified using laser-ablation inductively coupled plasma mass spectrometry (LA-ICP-MS), whereas the passage was monitored through high spectral imaging (HSI). Larger fluorescent signals were observed at the trichome tip compared to its base, indicating that trichomes served as a sink for the PSNP deposition in the plant system (Li et al. 2025).

2.5 Limitations and Knowledge Gaps

There are very few studies conducted so far on MNP in crop systems, and most of the studies done so far to elucidate absorption and adsorption are experimental. However, this is important to improve our understanding of the process of uptake, transport, and toxicity assessment. The analytical challenges and technological limitations needs to be overcomed. For example, issues of organic matter digestion, matrix interferences, spatial resolution under the microscope, limitations in spectral capabilities of FTIR, micro-Raman, LDIR, to identify smaller-sized particles, and of the Py-GCMS, ICP-MS, to provide concentration in terms of numbers and sizes limits application to real-world environmental samples. A quantitative understanding of MNPs under real-world conditions should be used to underpin the hazard and risk assessment of these particles to human and non-human biota. Nevertheless, the accurate quantification and visualization of MPs and NPs in higher plants remains an extensive challenge that remains sparingly explored at this point.

The majority of the experimental studies have used polystyrene nanoparticles in a hydroponic medium. Plants grow in a complex environmental matrix and are often exposed to various types of MNPs. Experimental studies on other types of MNPs are extremely challenging, yet desirable.

Characterization of MNPs absorbed from natural soil or atmosphere is difficult. To date, studies isolating and characterizing MNPs from natural environments have been negligible.

The risk assessment of MNP-embedded crops has not yet been elucidated. Crop diversity and environmental factors should also be taken into account to elucidate the risks posed by MNPs in crops (Table 2.1).

Table 2.1 Adsorption and absorption of MPs and NPs in crops

Species	Polymer type	Absorption	References
Vicia faba (faba bean)	100 nm PS beads	Root tip cells	Jiang et al. (2019)
Triticum aestivum (wheat) and *Lactuca sativa* (lettuce)	200 nm and 2.0 mm PS and 200 nm and 2.0 mm polymethylmethacrylate beads (PMMA)	Root stele, shoots	Li et al. (2020)
Triticum aestivum (wheat)	100 nm PS beads	Root and shoot xylem, leaf veins	Lian et al. (2021)
Cucumis sativus (cucumber)	100–700 nm PS	Not studied	Li et al. (2021a)
Lepidium sativum (cress)	50, 500, 4800 nm PS MNPs	–	Bosker et al. (2019)
Lactuca sativa (lettuce)	PSNPs	Absorption of NPs through stomata and translocation to roots	Lian et al. (2021)
Zea mays (maize)	Carboxyl-modified PSNPs and amino-modified PSNPs	Absorption via the stomatal opening	Sun et al. (2021)
Triticum aestivum (wheat) and Zea mays (maize)	Polypropylene particles (PP)	Accumulation in roots of wheat and stem of maize plant	Erdem et al. (2023)
Italian lettuce, radish, wheat, and corn	PSNPs (100 nm) and PSMPs (5 μm)	Absorption in roots	Gong et al. (2021)
Citrus root stocks	PSNPs (20 & 50 nm)	In rootstocks	Hussain et al. (2025)
Hordeum vulgare (barley), *Triticum aestivum (wheat), Daucus carota* (carrot), *Lactuca sativa* (lettuce)	PSMPs (500 nm)	MPs affecting lettuce and carrot	Zantis et al. (2023)
Triticum aestivum (wheat)	PSNPs (200 nm)	Wheat root xylem cell wall and in the xylem vessel to the shoots	Li et al. (2023b)
Arachis hypogaea (peanut), *Oryza sativa* (rice)	PS-NPs (80 nm)	PS-NP translocation from the root and accumulation in the grains at the maturation stage	Jiang et al. (2022)J

(continued)

Table 2.1 (continued)

Species	Polymer type	Absorption	References
Solanum lycopersicum (tomato)	PS-SO3H, PS, F-PS-SO3H (green fluorescently labelled), F-PS-NH2 (green fluorescently labelled), and PS-Eu NPs	Tomato leaves	Shi et al. (2023)
Oryza sativa (rice)	PSNPs (80 nm) and PSMPs (1 μm)	Root stele, stem vascular bundles and leaf veins, and mostly aggregated on cell walls and in the intercellular regions	Liu et al. (2022)
Oryza sativa (rice)	PS-COOH and PS-NH_2	Absorption of PS, PS-COOH, and PS-NH_2 in rice roots was confirmed by confocal microscope.	Wang et al. (2022)
Fragaria × *ananassa Duch. cv. 'Akihime'* (strawberry)	PS-MPs (100–200 nm)	PSMPs entered the roots and further translocated to the vascular bundles of petioles	Zhang et al. (2023)
Pisum sativum (pea)	PSNPs (20 nm)	PSNPs entered the roots and moved to the aerial parts	Kim et al. (2022)
Zea mays (maize) and *Glycine max* (soybean)	PSMNPs 80 nm and 500 nm	Larger sized remained on the leaf epidermis, whereas a few particles were present in the cytoplasm or apoplast. 80 nm migrated to the maize leaf interior and soybean leaves	Guo et al. (2023)
Gossypium sativum (cotton)	Red fluorescent-labelled PS-COOH (Shi et al. 2023; Li et al. 2019, 2024) and non-fluorescent PS-COOH with a particle size of 120 nm	PS-COOH invaded cotton roots	Li et al. (2024)
Lettuce	Green fluorescently labelled polystyrene microplastic microbeads with 200 nm diameter ($\lambda_{excitation}$ 488 nm; $\lambda_{emission}$ 525 nm)	Roots	Xu et al. (2022)
(*Triticum aestivum*) and lettuce (*Lactuca sativa*)	PSNPs 200 nm	Accumulation in roots, limited translocation in shoots	Luo et al. (2022)
Sweet potato	PE spheres (5 μm)	PE MP-fluorescent spheres were only observed in cortical tissues of roots and stems	Shi et al. (2022)
Zea mays (maize)	Small polystyrene (PS) beads (0.2, 0.5 and 1.0 μm) and large PS beads (2.0 and 5.0 μm)	More PS beads were absorbed by the root maturation zone than by the *root tip* zone	Li et al. (2023a)

(continued)

Table 2.1 (continued)

Species	Polymer type	Absorption	References
Cucumis sativus (cucumber)	Polystyrene (PS) and poly(methyl methacrylate) (PMMA) nanoplastics (100 nm)	Uptake by roots and their translocation to other tissues	Li et al. (2021b)
cowpea, flowering cabbage, rutabagas, and chieh-qua	Poly(vinylchloride) (PVC), poly(ethylene terephthalate) (PET), polyethylene (PE), and polyadiohexylenediamine (PA66)	Stems, leaves in *flowering* cabbage samples and peas and pods in cowpea samples	Ye et al. (2024)
Allium cepa (onion)	Red Polystyrene microspheres (50 nm)	PS spheres were internalized in different compartments of root tissue	Giorgetti et al. (2020)
Lactuca sativa	100 and 500 nm PSNPs	Absorption by roots and translocation to stems and leaves	Hua et al. (2024)
Daucus carota (carrot)	100 and 200 nm PSNPs	Absorption by roots and translocation to aerial parts	Dong et al. (2021)
Zea mays (maize)	PET and PS NP oligomers	Absorption of the proactively exposed non-labelled, fluorescently labelled, or europium-labelled plastic particles by maize (*Zea mays* L.) leaves through stomatal pathways, as well as their translocation to the vascular tissue through the apoplastic pathway, and accumulation in trichomes was identified	Li et al. (2025)
Colocasia esculenta (taro)	PSNPs (50 and 100 nm) PSMPs (200 and 500 nm)	PS particles of diverse sizes displayed differential root adsorption and penetration, with PS-NPs capable of penetrating the root epidermis, whilst PS-MPs were wholly excluded	Xu et al. (2025)

References

Alimi OS et al (2018) Microplastics and nanoplastics in aquatic environments: aggregation, deposition, and enhanced contaminant transport. Environ Sci Technol 52(4):1704–1724

Azeem I et al (2021) Uptake and accumulation of nano/microplastics in plants: a critical review. Nano 11(11):2935

Bandmann V, Homann U (2012) Clathrin-independent endocytosis contributes to uptake of glucose into BY-2 protoplasts. Plant J 70(4):578–584

Bandmann V et al (2012) Uptake of fluorescent nano beads into BY2-cells involves clathrin-dependent and clathrin-independent endocytosis. FEBS Lett 586(20):3626–3632

Bélanger J, Pilling D (2019) The state of the world's biodiversity for food and agriculture. FAO

Bosker T et al (2019) Microplastics accumulate on pores in seed capsule and delay germination and root growth of the terrestrial vascular plant Lepidium sativum. Chemosphere 226:774–781

Cole M et al (2014) Isolation of microplastics in biota-rich seawater samples and marine organisms. Sci Rep 4(1):4528

Dehghanian Z et al (2023) Micro (nano) plastics uptake, toxicity and detoxification in plants: challenges and prospects. Ecotoxicol Environ Saf 268:115676

Del Real AEP et al (2022) Assessing implications of nanoplastics exposure to plants with advanced nanometrology techniques. J Hazard Mater 430:128356

Dong Y et al (2021) Uptake of microplastics by carrots in presence of As (III): combined toxic effects. J Hazard Mater 411:125055

Erdem İÇ et al (2023) Combined approaches for detecting polypropylene microplastics in crop plants. J Environ Manag 347:119258

Etxeberria E et al (2006) Fluid phase endocytic uptake of artificial nano-spheres and fluorescent quantum dots by sycamore cultured cells: evidence for the distribution of solutes to different intracellular compartments. Plant Signal Behav 1(4):196–200

Giorgetti L et al (2020) Exploring the interaction between polystyrene nanoplastics and Allium cepa during germination: internalization in root cells, induction of toxicity and oxidative stress. Plant Physiol Biochem 149:170–177

Gong W et al (2021) Species-dependent response of food crops to polystyrene nanoplastics and microplastics. Sci Total Environ 796:148750

González-Melendi P et al (2008) Nanoparticles as smart treatment-delivery systems in plants: assessment of different techniques of microscopy for their visualization in plant tissues. Ann Bot 101(1):187–195

Guo S et al (2023) Foliar uptake and in-leaf translocation of micro (nano) plastics and their interaction with epicuticular wax. Environ Sci Nano 10(4):1126–1137

Habibi N et al (2022) Microplastics in the atmosphere: a review. J Environ Expo Assess 1(1):6

Hartmann NB et al (2019) Are we speaking the same language? Recommendations for a definition and categorization framework for plastic debris. ACS Publications

Horton AA et al (2017) Microplastics in freshwater and terrestrial environments: evaluating the current understanding to identify the knowledge gaps and future research priorities. Sci Total Environ 586:127–141

Hua Z et al (2023) The review of nanoplastics in plants: detection, analysis, uptake, migration and risk. TrAC Trends Anal Chem 158:116889

Hua Z et al (2024) Internalization, physiological responses and molecular mechanisms of lettuce to polystyrene microplastics of different sizes: validation of simulated soilless culture. J Hazard Mater 462:132710

Huang X-F et al (2014) Rhizosphere interactions: root exudates, microbes, and microbial communities. Botany 92(4):267–275

Hurley RR et al (2018) Validation of a method for extracting microplastics from complex, organic-rich, environmental matrices. Environ Sci Technol 52(13):7409–7417

Hussain K et al (2025) Impact of polystyrene nanoplastics on physiology, nutrient uptake, and root system architecture of aeroponically grown citrus plants. NanoImpact 37:100536

Iqbal B et al (2023) Impacts of soil microplastics on crops: a review. Appl Soil Ecol 181:104680

Jadhav B, Medyńska-Juraszek A (2024) Microplastic and nanoplastic in crops: possible adverse effects to crop production and contaminant transfer in the food chain. Plants 13(17):2526

Jiang X et al (2019) Ecotoxicity and genotoxicity of polystyrene microplastics on higher plant Vicia faba. Environ Pollut 250:831–838

Jiang M et al (2022) Evidence and impacts of nanoplastic accumulation on crop grains. Adv Sci 9(33):2202336

Khoury CK et al (2014) Increasing homogeneity in global food supplies and the implications for food security. Proc Natl Acad Sci 111(11):4001–4006

Khoury CK et al (2022a) The Plants That Feed the World: baseline data and metrics to inform strategies for the conservation and use of plant genetic resources for food and agriculture. Food & Agriculture Org

Khoury CK et al (2022b) Crop genetic erosion: understanding and responding to loss of crop diversity. New Phytol 233(1):84–118

Kim D et al (2022) Translocation and chronic effects of microplastics on pea plants (Pisum sativum) in copper-contaminated soil. J Hazard Mater 436:129194

Li L et al (2019) Uptake and accumulation of microplastics in an edible plant. Chin Sci Bull 64(9):928–934

Li L et al (2020) Effective uptake of submicrometre plastics by crop plants via a crack-entry mode. Nat Sustain 3(11):929–937

Li Z et al (2021a) The distribution and impact of polystyrene nanoplastics on cucumber plants. Environ Sci Pollut Res 28:16042–16053

Li C et al (2021b) Quantification of nanoplastic uptake in cucumber plants by pyrolysis gas chromatography/mass spectrometry. Environ Sci Technol Lett 8(8):633–638

Li H et al (2023a) Uptake and distribution of microplastics of different particle sizes in maize (Zea mays) seedling roots. Chemosphere 313:137491

Li R et al (2023b) Visual tracking of label-free microplastics in wheat seedlings and their effects on crop growth and physiology. J Hazard Mater 456:131675

Li W et al (2024) Uptake and effect of carboxyl-modified polystyrene microplastics on cotton plants. J Hazard Mater 466:133581

Li Y et al (2025) Leaf absorption contributes to accumulation of microplastics in plants. Nature 641:1–8

Lian J et al (2021) Foliar-applied polystyrene nanoplastics (PSNPs) reduce the growth and nutritional quality of lettuce (Lactuca sativa L.). Environ Pollut 280:116978

Liese B et al (2024) Uptake of microplastics and impacts on plant traits of savoy cabbage. Ecotoxicol Environ Saf 272:116086

Liu Y et al (2022) Uptake and translocation of nano/microplastics by rice seedlings: evidence from a hydroponic experiment. J Hazard Mater 421:126700

Loder MGJ et al (2017) Enzymatic purification of microplastics in environmental samples. Environ Sci Technol 51(24):14283–14292

Lozano YM et al (2021) Microplastic shape, polymer type, and concentration affect soil properties and plant biomass. Front Plant Sci 12:616645

Luo Y et al (2022) Quantitative tracing of uptake and transport of submicrometre plastics in crop plants using lanthanide chelates as a dual-functional tracer. Nat Nanotechnol 17(4):424–431

Mateos-Cárdenas A et al (2021) Adsorption, uptake and toxicity of micro-and nanoplastics: effects on terrestrial plants and aquatic macrophytes. Environ Pollut 284:117183

Pérez-Reverón R et al (2022) Microplastics in agricultural systems: analytical methodologies and effects on soil quality and crop yield. Agriculture 12(8):1162

Sandvig K et al (2011) Clathrin-independent endocytosis: mechanisms and function. Curr Opin Cell Biol 23(4):413–420

Shi L et al (2022) Impact of polyethylene on soil physicochemical properties and characteristics of sweet potato growth and polyethylene absorption. Chemosphere 302:134734

Shi R et al (2023) Toxicity mechanisms of nanoplastics on crop growth, interference of phyllosphere microbes, and evidence for foliar penetration and translocation. Environ Sci Technol 58(2):1010–1021

Shi X et al (2024) Impact of microplastics on plant physiology: a meta-analysis of dose, particle size, and crop type interactions in agricultural ecosystems. Sci Total Environ 955:177245

Sun X-D et al (2020) Differentially charged nanoplastics demonstrate distinct accumulation in Arabidopsis thaliana. Nat Nanotechnol 15(9):755–760

Sun H et al (2021) Foliar uptake and leaf-to-root translocation of nanoplastics with different coating charge in maize plants. J Hazard Mater 416:125854

Taylor SE et al (2020) Polystyrene nano-and microplastic accumulation at Arabidopsis and wheat root cap cells, but no evidence for uptake into roots. Environ Sci Nano 7(7):1942–1953

Uddin S et al (2022) Micro-Nano plastic in the aquatic environment: methodological problems and challenges. Animals 12(3):297

Wang J et al (2022) Effects of polystyrene nanoplastics with different functional groups on rice (Oryza sativa L.) seedlings: combined transcriptome, enzymology, and physiology. Sci Total Environ 834:155092

Xu Z et al (2022) Toxic effects of microplastics in plants depend more by their surface functional groups than just accumulation contents. Sci Total Environ 833:155097

Xu G et al (2025) When nano-and microplastics meet Taro (Colocasia esculenta) roots: their size-dependent adsorption, penetration, and promotion on secondary wall reinforcement. Environ Sci Technol 59:8345

Ye Q et al (2024) Identification and quantification of nanoplastics in different crops using pyrolysis gas chromatography-mass spectrometry. Chemosphere 354:141689

Yu Z et al (2024) Uptake and transport of micro/nanoplastics in terrestrial plants: detection, mechanisms, and influencing factors. Sci Total Environ 907:168155

Zantis LJ et al (2023) Species-dependent responses of crop plants to polystyrene microplastics. Environ Pollut 335:122243

Zhang C et al (2023) Potential translocation process and effects of polystyrene microplastics on strawberry seedlings. J Hazard Mater 449:131019

Chapter 3
Sample Collection, Preparation, and Identification of Microplastics in Soil: Issues and Recommendations

Montaha Behbehani, Saif Uddin, Yasmeen Al-Babtain, Maha Al-Sinan, and Shua'a Al-Ruwayeh

Abstract The omnipresence of microplastics (MPs) in agricultural soils, driven by practices such as compost and fertilizer addition, sewage sludge application, plastic mulching, and wastewater irrigation, is the main input of MPs to soil and poses significant risks to soil health and food safety. The paper evaluates methodologies from 39 global studies to identify gaps and inconsistencies in MP sampling, preparation, and identification protocols. Key findings reveal diverse sampling strategies, with composite sampling across depths (0–30 cm) and grid/quadrant approaches recommended to account for soil heterogeneity. Pre-treatment methods, including organic matter removal via 30% H_2O_2 or enzymatic digestion, and density separation using NaCl or NaI, vary widely, impacting MP recovery rates. Identification techniques predominantly combine microscopy with spectroscopic methods (FTIR, Raman), though challenges persist in false positives, polymer degradation, and nanoparticle detection. Methodological disparities in reporting units (e.g., MP/kg vs. mg/kg) hinder data comparability, complicating risk assessments and policy formulation. Quality assurance measures, such as procedural blanks, contamination controls, and spiked recoveries, are inconsistently applied, underscoring the need for standardised protocols. Recommendations emphasise harmonised sampling designs, optimised extraction workflows, and integration of advanced techniques like laser direct infrared spectroscopy for accurate quantification. Establishing global standards is imperative to generate reliable datasets, inform regulatory frameworks, and mitigate agricultural plastic pollution effectively.

Keywords FTIR · Raman · LDIR · Quality assurance · Risk assessments

M. Behbehani · S. Uddin (✉) · Y. Al-Babtain · M. Al-Sinan · S. Al-Ruwayeh
Environment and Life Sciences Research Centre, Kuwait Institute for Scientific Research, Safat, Kuwait
e-mail: sdin@kisr.edu.kw

S. Uddin, M. Zaman (eds.), *Micro- and Nano-plastics in Soil and Crop Systems*, Sustainability Sciences in Asia and Africa, https://doi.org/10.1007/978-981-95-2740-3_3

3.1 Introduction

The ubiquitous presence of microplastics in agricultural soils as a result of the use of sewage sludge, sludge-based fertilizers, composts, plastic mulch films, and wastewater for irrigation has emerged as a critical environmental concern for both soil health and food safety. For assessing the risks and generating intercomparable data across regions, harmonised and standardised methodologies are needed. Some efforts have been made (Withana et al. 2024; Uddin et al. 2020a) to harmonise methodologies for sampling, extraction, identification, and quantification that are essential to assess the level of microplastic contamination and ecological risks. Robust, harmonised methods for soil MP analysis are vital to enable cross-national data comparability and track progress toward environmental targets. Hartmann et al. (2019) emphasized the need to prioritise standardised protocols so countries can generate consistent and reliable data to map plastic contamination trends, inform policy decisions, and accelerate global mitigation efforts to reduce plastic pollution.

The global efforts to establish the legally binding plastic treaty underscore the need for standardised monitoring and reporting frameworks (UNEP 2025), as emphasized at the 5 United Nations Environment Assembly with the establishment of the Intergovernmental Negotiating Committee (INC). The first part (INC-5.1) of the fifth session took place from 25 November to 1 December 2024 in Busan, Republic of Korea. The second part (INC-5.2) took place from 5 to 14 August 2025 in Geneva, Switzerland.

In this chapter, we have attempted to critically evaluate the methods employed by various workers; 39 studies have been highlighted to identify the methodological differences and data quality objectives.

3.2 Sample Collection

Soil being a heterogeneous matrix requires appropriate considerations to select a representative site for microplastic determination. However, the criteria for sample site selection needs to be aligned with the research objectives (Withana et al. 2024; Uddin et al. 2020a, b, 2021, 2023; Kumar et al. 2020; Zaman and Uddin 2024).

Selecting an appropriate site for soil sampling requires a comprehensive assessment of the area's historical use, particularly practices like agricultural plastic mulching, municipal waste disposal, and irrigation with treated wastewater, as these directly influence MP presence. For a realistic assessment, both localised sources and dispersed origins should be evaluated. On large spatial scales, the topography, soil type, and land use-land cover are important considerations. The time of sample collection is also a critical factor; heavy rainfall can wash out the lighter MPs into the waterways (Schell et al. 2022). In agricultural soils where plastic mulch, sewage sludge, or compost is applied, sampling should be done before the next planting season. If the edge effect and carryover are not the objective of the study, samples should be taken centrally within fields (Piehl et al. 2018), whereas others have suggested that samples be taken on all four corners and from the center of a quadrant to make a composite sample (Yang et al. 2021a; Beriot et al. 2021). The ISO 18400-102 is designed for soil sampling (ISO 2017) and is applicable for MPs, employing random, stratified, or targeted approaches based on research goals (Möller et al. 2020).

3.2.1 Sampling Design

Based on the review of the sampling strategies used in the studies, it is observed that most of the studies adopted a grid-based sampling strategy or a quadrant system to ensure collection of a representative sample.

Several of these studies have adopted an approach of stratified sampling by collecting samples from different depths. The most commonly sampled depths were 0–10, 10–20, and 20–30 cm (Table 3.1). Most commonly the sampled layer is shallow (0–10 cm) it is suitable for surface accumulation from mulch fragmentation, sludge, and compost applications. Some studies, like that in Schleswig-Holstein, Germany, and Xinjiang, China, have sampled 10–30 cm as they intended to look at the vertical migration of MP in the soil profile. Few have looked at different layers within the profile, like the report from Shaanxi, China, where the samples were collected at 0–10, 10–20, and 20–30 cm to assess MP distribution.

While samples are collected, it should be kept in mind that most agricultural farms undergo tilling and furrowing that often disturbs the topsoil, and the presence of MPs might not solely be due to vertical migration.

3.2.2 Soil Collection

Most of the studies utilized a quadrant-based sampling approach, usually sampling the surface layer 0–10 cm using metal or stainless steel spoons, shovels, or scoops (Withana et al. 2024; Corradini et al. 2019; Crossman et al. 2020; Ding et al. 2020; He et al. 2021; Bello et al. 2025). However, to collect a deeper soil profile (30 cm or more), sediment corers, augers, or Lenz samplers are usually used (Möller et al. 2020; Weber et al. 2021). Most researchers have excluded plastic components to prevent cross-contamination, which is essential, so samples should be stored in metal or glass containers. Depending on the tools like Pürckhauer augers may necessitate combining multiple cores from adjacent sites to make a composite sample as larger sample size is needed for representative samples for MP in soils. Alternatively, larger pile-driving probes can be used, though they will be expensive (Weber et al. 2022).

In most studies, the surface microplastic concentration is higher than the deeper layer. Liu et al. (2018) reported 78.00 ± 12.91 MP kg^{-1} in surface soil and

Table 3.1 Summary of the methodology adapted for microplastics in soil by different workers

Country/ region	Sampling			Sludge Samples		Sample extraction				Identification	Units reported	Quality control	
	Matrix	Sampling depth	Pre-treatment	Matrix	Treatment	Pre-digestion	Flotation	Filtration	Post-oxidation				Reference
Nanjing and Wusi Region, China	*Soil Samples:* Soil was collected from agricultural lands, with no history of sludge application and treated wastewater use.	Surface soils were collected using a stainless steel shovel.	Samples were air-dried and sieved through 5-mm and 1-mm stainless steel sieves. Microplastics larger than 1 mm were manually removed from the sieve.	Collected from 3 WWTPs	Samples were air-dried and sieved through 5-mm and 1-mm stainless steel sieves. Microplastics larger than 1 mm were manually removed from the sieve.	30% H_2O_2 at 70 °C to remove organic matter (OM). Samples heated until nearly dry.	Solutions tested: NaCl (1.2 g/cm^3), $ZnCl_2$ (1.5 g/cm^3), NaI (1.8 g/cm^3). Shaken for 1 hr (initial) and 30 min (repeat), settled for 48 hr. Supernatant siphoned and combined.	Vacuum filtration using 20 μm pore size nylon membrane filter. PTFE and glass fiber membranes were avoided due to shedding.	MPs washed with **30% H_2O_2 + H_2SO_4 (3:1 v/v)** at 70 °C to remove residual OM.	Stereo microscope: Observed shape (fiber, fragment, bulk), color, and size (0.02–5 mm). ImageJ is used for size classification. *μ-FTIR (Micro-Fourier transform infrared spectroscopy):* ATR mode, 4 cm^{-1} resolution, 32 scans. Spectra were compared with those in the Hummel Polymer and Additives databases. Three random 1x1 cm^2 areas were analyzed per membrane.	MP abundance: items/kg (soil or sludge).	*Contamination prevention:* Glassware only, rinsed with ultrapure water. Researchers wore cotton lab coats. Flasks covered with aluminum foil during flotation. *Membrane selection:* Tested quartz, glass fiber, PTFE, and nylon membranes; selected nylon to avoid fiber shedding. *Replicates:* Three parallels per treatment. *Oxidation validation:* Tested virgin polymers (PE, PP, PS, PA, ABS, PET) to ensure no degradation during oxidation. *FTIR verification:* Non-plastic particles (e.g., black carbon) were excluded via spectral analysis.	Li et al. (2019)

Shanghai, China	20 vegetable farmlands were samples GPS coordinates recorded for each site.	0–3 cm shallow layer and 3–6 cm deeper layers, measured with calipers.	1 kg of soil per sample, visible debris >5 cm removed Stored in aluminum boxes to avoid plastic contamination.			Dried soil at 70 °C for 24 hours.	Density separation using NaCl (1.19 g/ cm^3) and ultrasonicated for 2 minutes. Stirred for 30 minutes, settled for 24 hours.	Filtered through 20 µm nylon membrane filter. Repeated 3 times for complete extraction.	Treated supernatant with 30% H_2O_2 at 50 °C for 72 hours	Visual classification using stereomicroscope (Nikon SMZ25) for shape (fiber, fragment, film, pellet) and color. Size categories: microplastics (<5 mm), mesoplastics (5 mm–2 cm). Polymers identified using µ-FTIR In transmittance mode (4000–675 cm^{-1}, 32 scans). Spectra were matched to Hummel Polymer database (≥70% match threshold). Ninety-nine randomly selected particles were analyzed.	Microplastic/ mesoplastic abundance in items kg^{-1} of dry soil.	ForrRecovery test samples were spiked with PP, PE, PA, PET, PVC, PC, ABS, PMMA, and PS microplastics. Recovery rates were > 90% for PP, PE, PA, PC, ABS, PMMA, and PS; PET and PVC not recovered due to its density higher than NaCl. Glass/metal equipment rinsed with Milli-Q water. No plastic containers used. Researchers wore cotton lab coats and gloves. Samples stored in closed containers to avoid airborne contamination. Triplicate blank tests confirmed no microplastics in controls.	Liu et al. (2018)

(continued)

Table 3.1 (continued)

Country/ region	Sampling			Sludge Samples		Sample extraction				Identification	Units reported	Quality control	
	Matrix	Sampling depth	Pre-treatment	Matrix	Treatment	Pre-digestion	Flotation	Filtration	Post-oxidation				Reference
Shangsuan, Anle, Dunshang, Dagoujian, and a buffer zone, China	Five agricultural soils in the Dian Lake basin	Collected six sub-samples per plot using a narrow spade from two depths 0–5 cm and 5–10 cm	Composite samples were sieved through a 10 mm mesh. Air-dried at 25 °C, roots and large debris removed.	Composite agricultural soil	Wet-sieving of air-dried soil through 2.00, 0.25, and 0.05 mm sieves. After that, the samples were centrifuged for 10 minutes at 2300 rpm to separate aggregates. The >2 mm were considered as large macro-aggregates, 0.25–2 mm as small macro-aggregates, 0.05–0.25 mm as micro-aggregates.	Samples treated with 35% H_2O_2 and 10% $FeSO_4$ at 50 °C to remove organic matter	Saturated NaI with a density of 1.8 g/cm^3 solution for density separation.	Supernatant filtered through stainless steel meshes (1, 0.25, 0.05 mm).		Visual classification using dissection microscope at 40× Categorized into size range of 10–1 mm, 1–0.25 mm, 0.25–0.05 mm and shapes were identified as fiber, fragment, and film, while pellets and foams were excluded	No polymeric characterisation as done.		Zhang and Liu (2018)

Xinjiang, China. Quadrat Sampling	Composite samples from a quadrat measuring 200 cm × 100 cm × 30 cm depth from 593 sites were collected.	Soil samples were manually collected from the top 30 cm.	A questionnaire was sent to 672 farmers, 593 responded giving information of crop type, mulch film thickness, mulching years, etc.	Ultrasonic cleaning to remove soil particles.	Blotting paper is used to absorb moisture. The samples were then air-dried and weighed.					Visual identification was done and no spectroscopic methods were used.	The results were reported as film abundance in kg ha^{-1} for individual sites and in Metric tons for the regional total	Large sample size to reduce bias. Ultrasonic cleaning to ensure accurate weight measurements. Compared results with national standards (GB/T 25413-2010). But no mention of procedural blanks or contamination controls during lab analysis. And potential self-reporting bias in questionnaire data.	Zhang et al. (2016)

(continued)

Table 3.1 (continued)

Country/ region	Sampling			Sludge Samples		Sample extraction				Identification	Units reported	Quality control	
	Matrix	Sampling depth	Pre-treatment	Matrix	Treatment	Pre-digestion	Flotation	Filtration	Post-oxidation				Reference
Szeged, Hungary	Soil samples from greenhouse and farmlands. The area was divided into 15 areas of 52 × 9 m and Three areas were randomly selected for sampling.	Soil samples were collected from two depths, i.e., 0–20 cm and 20–40 cm using metallic auger	Soil samples were oven-dried, homogenized, sieved and composite samples were used. Visible macroplastic debris on the surface was hand-picked by two observers. The collected plastics were precleaned and stored in large polyethene bags.			Organic matter was digested using 30% H_2O_2 and Fenton reagent	$ZnCl_2$ was used for density separation.	Extracts were filtered through nylon and Whatman filters, dried. Macroplastic materials were soaked in tap water for 48 hours, washed, sieved, dried, separated, counted, and measured.		Microscopic examination of filters. Suspected microplastic particles were confirmed using the hot needle and Raman spectroscopic analysis.	Macroplastics were reported in kg ha^{-1} and categorized by size (0.5–1.0 cm, 1–5 cm, 5–10 cm, 10–15 cm, >15 cm). Microplastics were reported as MP kg^{-1}	Cleaning the auger before each sampling, avoiding sample mix, and using minimal plastic materials. Rinsing laboratory apparatus with distilled water three times, wearing cotton lab coats and gloves, and covering samples with aluminum foil to prevent contamination.	Sa'adu and Farsang (2022)

Shenyang Agricultural University, China	Arable land	Soil cores were taken from random locations in each plot using a coring drill of 8.5 cm diameter from 0–10 cm, 10–20 cm, 20–40 cm, 40–60 cm, 60–80 cm, and 80–100 cm.	Soil samples were sieved through a 5-mm metal mesh to remove macroplastic residues and roots, and air-dried.			20 g of air dried soil samples was mixed with density separation solution, sonicated, filtered, and treated with 30% H_2O_2 solution, heated, and filtered again..	$ZnCl_2$ solution was used for density separation	The supernatant was filtered through a membrane filter, dried		Stereo microscope used for identification. The polymeric characterisation was done using μ-FTIR. Microplastics were classified into films, fibers, and granules.Size, and color were also reported	MP $m^{-2}$for macroplastics and MP kg^{-1} of soil for microplastics	Soil stored in cotton bags. Metal and glass was used and washed with pure water, and covered with aluminum foil. All the solutions were filtered, and cotton lab coats were worn. Three replicates without soil were run concurrently to check for potential plastic contamination during extraction. No plastics were detected in these blank control samples. Positive control samples were used to test the efficiency of the extraction method, achieving recovery rates of 87% ± 3% for film microplastic and 90% ± 7% for fiber microplastic.	Li et al. (2022)

(continued)

Table 3.1 (continued)

Country/ region	Sampling			Sludge Samples		Sample extraction				Identification	Units reported	Quality control	
	Matrix	Sampling depth	Pre-treatment	Matrix	Treatment	Pre-digestion	Flotation	Filtration	Post-oxidation				Reference
China	Experimental soil where PS was added		Seeds were placed in sterilized beakers with PS-MP solutions and grown in a climate-controlled atmosphere for 14 days at 25 ± 1 °C and 75% humidity.							Transmission electron microscopy (TEM) and scanning electron microscopy (SEM) were used to observe PS-MP distribution in rice seeds, roots, and buds.	mg L^{-1}	Samples were fixed, rinsed, dried, and coated for SEM observation. TEM samples were embedded, sliced, and dispersed on copper mesh.	Zhang et al. (2021)
Changsha, China	Agricultural soil	0–20 cm	The soil was air-dried, debris removed and passed through a 10 mm steel sieve, and stored in dark conditions										Yang et al. (2021a)

Ishikawa Prefecture, Japan	19 paddy fields in the Tedori River alluvial fan, Five points per field (four corners and centre) using a 15-cm-diameter core sampler	0–15 cm, divided into two layers (0–10 cm and 10–15 cm). 1 kg soil sample was also taken from the centre for physico-chemical analysis.	The soil samples were dried at 60 °C for 2 days			Wet Sieving was performed by mixing with tap water, stirring, settling overnight, sieved using 1-mm mesh.		Sieved through 1 mm sieve	Visual Sorting of Microcapsules based on the criteria: 2–3 mm. and color White/ yellow discoloration, spherical hollow capsules.	SEM used for analyses of surface morphology of microcapsules. FT-IR for polymer identification	Reported in mg/ kg soil and g/m². The results were also reported as pieces/kg soil and pieces/m².	Two additional individuals verified visual sorting to minimize errors. Spiked samples were used and 85–95 recovery	Katsumi et al. (2021)
Tainan City, Taiwan	Five farmlands (A–E) with different crops and mulch. Each farm is divided into three subplots; composite samples collected from, five subsamples per subplot	0–5 cm top layer and 20–25 cm bottom layer	Air-dried, homogenized, and sieved using 1 and 5 mm stainless-steel sieves			30% H_2O_2 used overnight to degrade organic material	NaCl solution, stirred, settled for 2 hours, and	Filtered through 1 µm glass microfiber filters.	Stereomicroscope used to classify MPs by shape (fibers, fragments, films, foams, pellets, microbeads) and color.	ATR-FTIR used for polymeric identification	MP/m²	Control farm as a background control. Control Soil sample ignited at 500 °C to remove impurities tested for contamination during transportation. Spiked samples were also analyzed	Fakour et al. (2021)

(continued)

Table 3.1 (continued)

Country/ region	Sampling			Sludge Samples		Sample extraction				Identification	Units reported	Quality control	
	Matrix	Sampling depth	Pre-treatment	Matrix	Treatment	Pre-digestion	Flotation	Filtration	Post-oxidation				Reference
Sulur, Nagondapalli, Royakottah, Krishnagiri regions, Tamil Nadu, India	Agricultural soils from areas that have been mulched for 2–10 years	0–10 cm, 11–20 cm, and 21–30 cm.	A stainless steel probe was used to collect 15 soil cores, composited. Control samples (virgin soil) taken 500 m away		The equipment was cleaned with 70% ethanol. Samples were air-dried, powdered, and stored at 4 °C. Soil sieved using stainless steel sieves (pore sizes: <5 mm to <1 mm).	Digested with 30% H_2O_2 at 50 °C for 72 hours.	Floating organic matter separated via saturated NaCl solution.	Filtered, dried, and plastic weight calculated as: $W = w\_1 - w\_2$ (initial and final filter weights)		Stereo microscopy, shapes identified as films, fibers. Further identification of surface morphology and elemental composition was done using SEM-EDX	g kg^{-1} soil and also MP/kg soil	Glass containers, nitrile gloves, and cotton lab coats were used. Virgin soil samples for baseline comparison. Five replicates for enzyme activity assays. The equipment was sterilized with 70% ethanol between samples.	Kumar and Sheela (2021)
Yeoju City, South Korea	100 sites covering forest ($n = 8$), traffic/roadside ($n = 19$), residential ($n = 5$), and agricultural ($n = 68$) areas. Subdivided into orchard, paddy field, upland, and greenhouse. Systematic 2.5 km × 2.5 km grid.	0–5 cm.	Hand auger used to collect three replicates per site. Samples stored in stainless steel containers. Oven-dried at 60 °C for 48 hours. Mesoplastics separated using a 5-mm sieve.			Treated with $FeSO_4$, H_2SO_4, and H_2O_2, heated at 75 °C for 24 hours	Microplastics (<5 mm) extracted using $ZnCl_2$ solution (1.7 g cm^{-3}), stirred, settled, and repeated thrice.	Filtered through 0.45 μm cellulose acetate membranes		Digital stereo microscope used to count and classify by shape (fragment, film, fiber, sphere) and color (black, red, green, etc.). Polymers identified using FTIR by matching spectra to reference databases.	MP kg^{-1} soil, categorized by size: <1 mm, 1–5 mm, >5 mm and Polymer Types reported as percentages (e.g., SBR: 43%, PP: 22%).	Stainless steel tools and containers used. Triplicate samples per site. FT-IR analysis on 60 representative samples to confirm polymer composition.	Choi et al. (2021)

Swiss flood plain, Switzerland	Remote high mountain soil and from populated regions	Collected from topsoil layers (depth unspecified) using standardized soil sampling protocols. Mixed samples from each site were analyzed (three replicates per site).			Custom-built setup involving a vacuum pump, magnetic mixer, and aeration system	Tested HNO_3 (90 °C, 48 h) for organic matter digestion	Combined with centrifugation; rubber discs are used to prevent resuspension of mineral particles.	Particles filtered through anodisc membranes (0.45 µm porosity).		Polymer identification using Raman spectroscopy (e.g., PP, PE, PVC, PET). FT-IR validation.	MP/kg^{-1} of soil, and categorized by size as small microplastics (<1 mm), Large microplastics (1–5 mm), Mesoplastics (>5 mm)	Anodisc filter to rule out contamination. Recovery tests were conducted. Triplicate samples per site. Chemical resistance testing after HNO_3 treatment. Negative controls.	Scheurer and Bigalke (2018)

(continued)

Table 3.1 (continued)

Country/ region	Sampling			Sludge Samples		Sample extraction				Identification	Units reported	Quality control	
	Matrix	Sampling depth	Pre-treatment	Matrix	Treatment	Pre-digestion	Flotation	Filtration	Post-oxidation				Reference
Swiss Seeland, Switzerland	11 agricultural drainage sites (8 passive flow systems, 3 basin systems)	Flowing water was directly collected into 3 L pre-cleaned glass containers. Basin water was sampled with a pre-cleaned metal bucket, transferred to glass containers. 5 sites sampled to assess temperal variability.				Backwashed with 15 mL 60% H_2SO_4 for 30 minutes. Filtered through 8 μm cellulose nitrate membrane filters	Centrifugation (3000 RPM, 30 min) to isolate MP	Supernatant filtered through 0.2 μm Anodisc filters for FTIR analysis		FTIR spectroscopy, particles >100 μm analyzed.	MP L^{-1} with mean ± standard deviation.	Processed 2.5 L Milli-Q® water; as a blank with mean blank = 0.8 MP L^{-1}. Detection limit: Mean blank +3SD = 2.2 particles L^{-1}. Recovery test with 100% recovery of spiked PE, PP, and PVC particles (1–2 mm). One sample in triplicate, one in duplicate; mean RSD = 54%. Temporal replicates (same site, different dates): mean RSD = 72%. For minimizing contamination lab equipment rinsed with Milli-Q® and ethanol; sample preparation under HEPA-filtered laminar flow hood.	Bigalke et al. (2022)

Speyer, Germany	Experimental agricultural farms where 190 T of sludge per hectare was applied										MP g^{-1}		Tagg et al. (2022)
Ilia County, Greece	Agricultural fields and Five uncultivated "blank" samples	0–30 cm using stainless-steel shovel	Air-dried, sieved through 5 mm mesh, stored in aluminum containers			Treated with 30% H_2O_2 at 70 °C for 2 days to remove organic matter.	Saturated NaCl (1.18 g/cm³) mixed with 150 g soil stirred, rested, and filtered through 500 μm and 20 μm sieves. Repeated until no floating material			Stereo microscope to identify 20 μm–5 mm. ATR-FTIR to identify polymer type. SEM for analyzing surface degradation of MPs.	MP kg^{-1} classified into 2 classes 20 to 500 μm 500 μm to <5 mm	Blank samples. Triplicates for each sample; Stainless-steel tools, aluminum containers, visual exclusion of non-BMF MPs. To avoid contamination.	Isari et al. (2021)
Murcia, Spain	3 fields sampled that had a Long-term use of LDPE, PAC, and biodegradable mulch over 12 years	0–10 cm and 10–30 cm depths			10 g soil or 5 g compost mixed with water, centrifuged (3000 rpm, 10 min), filtered through 2.5 μm filter.	Treated with H_2O_2 and heated (130 °C) to differentiate MPs from organics.				Visual examination using a Stereo microscope. Particles are classified by shape, color, and heat response. ImageJ software is used to quantify particle count and size (30 μm–2 mm). FTIR spectroscopy to identify polymers	MP kg^{-1} and $mm^2\ kg^{-1}$ (particle area).	Filed blank samples; Three replicates per sample. Instrument Validation. Detects only light-density MPs >30 μm. Potential underestimation of non-floating plastics	van Schothorst et al. (2021)

(continued)

Table 3.1 (continued)

Country/ region	Sampling			Sludge Samples		Sample extraction				Identification	Units reported	Quality control	
	Matrix	Sampling depth	Pre-treatment	Matrix	Treatment	Pre-digestion	Flotation	Filtration	Post-oxidation				Reference
Netherlands	2 farms with loamy sand, 7 years of compost; clay, 20 years of compost). Each farm had 3 fields, 5 points per field	0–10 cm and 10–30 cm depths		Municipal organic waste and garden/ greenhouse waste	10 g soil or 5 g compost mixed with water, centrifuged (3000 rpm, 10 min), filtered through 2.5 μm filter.	Treated with $H_2O_2$and heated (130 °C) to differentiate MPs from organics.				Visual examination using a Stereo microscope. Particles are classified by shape, color, and heat response. ImageJ software is used to quantify particle count and size (30 μm–2 mm). FTIR spectroscopy to identify polymers	MP kg^{-1} and $mm^2\ kg^{-1}$ (particle area).	Filed blank samples; Three replicates per sample. Instrument Validation. Detects only light-density MPs >30 μm. Potential underestimation of non-floating plastics	van Schothorst et al. (2021)
Winterthur, Switzerland	Processed organic waste from municipalities, gardens, and households	Samples at three stages: incoming organic waste (conveyor belt), sieving residues, and final solid digestate.	Sampling was conducted over five days (June–August 2017), 15 samples total (4.5 L each, collected every 40 minutes over 2 hours).		Bulk density measured for compost and digestate samples.	Washed samples under running water over a 6-mm mesh to separate plastics. Dried retained material at 40 °C for 4 days.				Visually inspected dried samples to collect plastic pieces >1 mm. Microscopic examination and classification into plastic bags, packaging materials, etc. FTIR spectroscopy for identification of polymers.	Tonnes yr^{-1} for material flow analysis. Plastic Concentrations as a percentage of wet organic waste in % and total accumulation in Swiss soils reported in tonnes	Monte Carlo iterations to quantify variability and uncertainty in plastic content	Kawecki et al. (2021)
Arusha, Tanzania	Soils from irrigated farms									ATR-FTIR			Kundu et al. (2022)

Mauritius	Agricultural soils collected from three sites	0–10 cm and 10–20 cm	Each site is divided into a 20 × 20 m quardant. 40 samples per grid composited into a single sample. Samples were dried, sieved, and analyzed	Grab sampling from three WWTPs. Sludge (1 L) was collected after primary/ secondary treatment; wastewater (500 mL) was collected before discharge. Samples were stored in glass bottles.	Sludge samples were mixed with water, NaCl, centrifuged, and digested with H_2O_2. The effluents were directly mixed with H_2O_2, digested, and filtered.	H_2O_2 digestion (7 days) to remove organic matter.	Flotation with distilled water and NaCl (1.19 g/cm³), followed by centrifugation	Filtered through 0.45 μm cellulose nitrate membranes.		Microscopic examination to categorize MPs by shape (fiber, fragments, flakes, films, spheres), color, and size (0.25–5 mm). ATR-FTIR for polymer identification of particles >0.5 mm.	MP kg^{-1} for soil. MP L^{-1} for effluent MP kg^{-1} wet weight for sludge	Spiked samples with known plastics to calculate recovery rates; three replicates. Blank samples analyzed. Lab equipment rinsed with ethanol; use of gloves and aluminum foil used to cover glassware.	Ragoobur et al. (2021)
Tunisia	4 Agriculture soils with organic farming, control; Mulched; Greenhouseand soil irrigated with treated wastewater.	10–20 cm depth	Earthworms (*Eisenia andrei*) were exposed to soil spiked with MPs (<100 μm) for 7 and 14 days.			MPs in earthworms were digested using 10% KOH.	Density separation using saturated NaCl (1.19 g.cm⁻³). The earthworm samples were density-separated by KI	Sieved through 300 μm mesh to retain particles (1 mm–300 μm). Earthworm samples were filtered through 0.22–3 μm cellulose nitrate filters.		Microscopic examinatio for color/shape. FTIR and Raman microspectroscopy for polymer identification	MP kg^{-1} in soil. MP g^{-1} tissue for earthworm	Rinsing equipment with Mill-Q water. Blank samples. 10 replicates	Boughattas et al. (2021)

(continued)

Table 3.1 (continued)

Country/region	Sampling			Sludge Samples		Sample extraction				Identification	Units reported	Quality control	
	Matrix	Sampling depth	Pre-treatment	Matrix	Treatment	Pre-digestion	Flotation	Filtration	Post-oxidation				Reference
Chile	Agricultural soil with varying degrees of sludge application histories (0–5 applications over 10 years) from 31 fields and a Control site (no sludge) applied	0–25 cm sampled with a metallic auger; three replicates per field	Dried at 40 °C, sieved with <2 mm sieve, and stored in PET jars	Seven samples collected from fields undergoing sludge applications during the study	Dried, milled, sieved (<1 mm), and stored I PET jars		Sequential use of NaCl (5 M, $\rho = 1.20$ g cm^{-3}) and $ZnCl_2$ (5 M, $\rho = 1.55$ g cm^{-3}) solutions to float microplastics	5 g of soil and 1 g of sludge is mixed with deionized water, stirred, centrifuged, and filtered.	Supernatants filtered through Whatman No. 42 filters (retention >8 µm).	Microscopic visual identification and classified into fibers, fragments, films, or pellets based on morphology. No spectroscopic confirmation was done.	MP g^{-1} in soil/Sludge (DW). MP weight estimated using particle dimensions and assumed polymer densities.	Reagent blanks were included in each batch to detect lab contamination. Ignited soil samples tested for transport/storage contamination and recovery test. Five samples were analyzed five times; the coefficient of variation = 12%.	Corradini et al. (2019)

Ontario, Canada	Soil from four agricultural fields (three biosolid-treated, one control)	0–5 cm, 5–10 cm, and 10–15 cm using a stainless steel corer (5 cm diameter)	Soil cores wrapped in aluminum foil, frozen, then oven-dried (≤65 °C); biosolids refrigerated before drying	Biosolidswere collected directly from application trucks/hoses during spreading		Fenton's reagent used to remove organic content.	Sequential use of sodium tungstate (Na_2WO_4) solutions to float microplastics (MPs)	Isolated MPs filtered onto glass fiber filters		Microscopic examination to classify MPs by shape. Polymer identification using ATR-FTIR for particles >300 μm and μ-FTIR for smaller particles/ fibers.	MP kg^{-1} for soil and Biosolids reported in MP kg^{-1} and MP L^{-1}. Mass estimationusing particle dimensions (e.g., fragment volume as ellipsoids, fiber volume as cylinders.	Aluminum foil covers during drying. Spiked samples (PE, PVC, PET fibers) showed recovery rates of 80–97%. Lab coats and controlled airflow in processing areas. Procedural blanks were analyzed to account for background contamination. Triplicate analysis of composite soil samples; standard errors reported.	Crossman et al. (2020)
Mexico	Arable land												Hernández-Gutiérrez et al. (2021)
Shaanxi, China	Agricultural soils	0–10 cm	6 composite samples per site, each composed of 3 sub-samples collected 3 m apart to a 2 kg quantity. Stored in clean metal bottles and refrigerated		The soil was oven-dried at 70 °C for 24 hours.	H_2O_2 was used for organic matter digestion. Heated at 65 °C, 12 hours.	50 g soil mixed with NaCl (1.2 g/cm^3) and $CaCl_2$ (1.5 g/ cm^3) solutions. The supernatant was collected after 24 hours.	Filtered through 0.45 μm filter paper, dried, and stored for analysis.		Microscopic examination to identify type, size, and shape. Followed by FTIR spectroscopy for polymer identification SEM for surface characterization.	MP kg^{-1}	Glass/metal tools used. Equipments rinsed with ultrapure water and samples wrapped to avoid airborne fibers. Blank experiments confirmed no MPs in controls.	Ding et al. (2020)

(continued)

Table 3.1 (continued)

Country/region	Sampling			Sludge Samples		Sample extraction				Identification	Units reported	Quality control	
	Matrix	Sampling depth	Pre-treatment	Matrix	Treatment	Pre-digestion	Flotation	Filtration	Post-oxidation				Reference
Tibetan Plateau, China	17 farms	(0–3 cm and 3–6 cm layers)	Visible debris (>3 cm) was removed; samples were stored in aluminum foil bags		Sediment/soil were oven-dried (70 °C, 24 h), sieved (2 mm)	Digested with 30% H_2O_2 and 0.05 M Fe(II) solution at 75 °C for 72 hours to degrade organic matter.	Density separation using saturated NaCl (1.2 g/cm^3). Supernatants collected three times for thorough extraction.	Filtered through 0.45 μm membranes		Microscopic examination to classify MPs by shape, size, and color. Raman spectroscopy for polymer composition	MP Kg^{-1}	Tools rinsed with ultrapure water; glassware heated at 400 °C; laminar flow cabinets used during experiments. Triplicate Milli-Q water blanks analyzed.	Feng et al. (2020)
Schleswig-Holstein, Northern Germany	Farmland 15 sites	0–10 cm, 10–20 cm, and 20–30 cm.	Boreholes were drilled at four corner points of a 40 × 40 m square per site. 540 soil samples (36 per site) totaling ~123 kg of soil. Stored in glass containers to avoid contamination		Soil soaked in water, homogenized and sieved through a 1 mm steel sieve to retain particles >1 mm.	For organic matter removal the aggregates dissolved manually; excess water removed with syringes				Sieved material was examined under a stereomicroscope for MP (1–5 mm) and classified by shape and color. FTIR spectroscopy for polymer composition.	MP kg^{-1}	Glass/metal tools used; lab coats worn; instruments rinsed between samples. Spiked soil samples used for recovery assessment and blank samples analyzed.	Harms et al. (2021)

Xinjiang Uygur Region, China	Agriculture fields with cotton plantation and 5, 15, and 24 years of mulching.	0–5, 5–20, 20–40 cm	Five 1 m × 1 m plots per field, triplicate soil cores at different depths and stored in aluminum bags			Digested with H_2O_2	Density separation using NaI (1.85 g/cm³). Air-dried soil mixed with NaI, agitated, settled, and supernatants collected. Flotation was repeated thrice.	Filtered through 7 μm glass fiber filters,					Huang et al. (2020)
Schleswig-Holstein, Northern Germany	Farmland boreholes were drilled at four corners of a 40 × 40 m agriculture fields.	0–10, 10–20, and 20–30 cm.	Soil soaked in water homogenized and sieved through a 1 mm steel sieve to retain particles >1 mm.			Organic matter removal was done by manually dissolving the aggregates and excess water removed with syringes.				Sieved material was examined under a stereomicroscope for MP (1–5 mm) and classified by shape and color. The polymers were confirmed using an FTIR spectrometer.	MP kg^{-1}	Stored in glass containers to avoid contamination. Glass/metal tools used; lab coats worn; instruments rinsed between samples. Spiked samples for recovery calculations. Blank samples were also analyzed.	Piehl et al. (2018)
Shouguang, Shandong and Xuzhou, Jiangsu, China.	Agricultural soils	0–10 cm	Plant debris was removed, air-dried, ground, and sieved with a 1 mm sieve.			Digestion with 30% H_2O_2 at 70 °C to remove organic matter.	NaCl (1.2 g/cm³) was used for density separation.	Filtration through 20 μm nylon membranes		Identification of polymers using laser direct infrared (LDIR) spectroscopy in the size range of 20–1000 μm.	MP kg^{-1}	Avoided plastic tools; cotton lab coats worn; glassware pre-washed with HNO_3, acetone, or hexane. Blanks processed. Triplicate extraction.	Li et al. (2021)

(continued)

Table 3.1 (continued)

Country/ region	Sampling			Sludge Samples		Sample extraction				Identification	Units reported	Quality control	
	Matrix	Sampling depth	Pre-treatment	Matrix	Treatment	Pre-digestion	Flotation	Filtration	Post-oxidation				Reference
Yingstan, Jiangxi Province Southeast China	Agricultural plots with long-term repeated pig manure application	0–20 and 20–40 cm	Plant debris removed; samples air-dried, homogenized, and sieved (e.g., ≤5 mm or ≤1 mm)			30% H_2O_2 to degrade organic material.	NaCl saturated solution for density separation	Filtration through a 0.45 μm glass fiber filter		A stereo microscope was used to count and categorise particle morphology and polymers were identified by using FTIR.	MP kg^{-1}		Yang et al. (2021b)
Murcia, Spain	6 vegetable fields	0–10 cm	Each field had 3 parcels, with 3 sampling points per parcel.	Sheep faces was collected	Samples were air-dried.	High organic matter in faeces hindered complete particle recovery.	Density separation using distilled water	Soil was shaken, centrifuged, and supernatant filtered. Ultrasonic bath used to break aggregates.		Microscopic examination and images magnification). Images were taken before/after heating (130 °C for 10 s) to identify plastics via shape, color, brightness, and heat response.	MP Kg^{-1}	Using stainless steel tools. Blank samples reported no contamination. However, no recovery tests performed.	Beriot et al. (2021)
Youg-In Province, Korea	Agricultural soil										MP Kg^{-1}		Kim et al. (2021)

Paotai Town, Shihezi City, Xinjiang, China.	Agricultural soil with varying period of mulching 0, 5, 15, and 30 years.	0–20 cm	Five samples per site were collected using a multipoint mixed method from a 0.1 × 0.1 m. The samples were air-dried, sieved through meshes (2 mm, 0.9 mm, 0.28 mm, 0.15 mm), and homogenized			98% H_2SO_4 treatment to oxidize organic debris.	Water was used to disintegrate soil aggregates and density separated by distilled water.	Vacuum filtration with quantitative filter paper.	Electrostatic separation by rubbing glass rods against silk to collect small particles.	SEM was used to identify MPs and ATR-FTIR used for polymeric identification	Mg kg^{-1}	Recovery rates tested with spiked samples. Stainless steel/ glass tools, lab coats, gloves, deionized water rinsing. Unmulched soil samples processed as control and Triplicate experiments for extraction efficiency.	Li et al. (2020)

(continued)

Table 3.1 (continued)

Country/ region	Sampling			Sludge Samples		Sample extraction				Identification	Units reported	Quality control	
	Matrix	Sampling depth	Pre-treatment	Matrix	Treatment	Pre-digestion	Flotation	Filtration	Post-oxidation				Reference
Gansu and Xinjiang, Northwest China:	Farmland with varying degrees of mulching	0–10, 10–20, and 20–30 cm.	Collected manually from three quadrats (100 cm × 50 cm × 30 cm) per field, divided - Microplastics (MiPs):** Soil samples (1 kg per layer) collected from pit sides, homogenized, air-dried, and sieved (2 mm).				Extracted via flotation in distilled water, centrifugation (14,400 × g, 5 cycles)	filtration (<3 μm filter), and thermal treatment (130 °C heating to melt LDPE particles)		Microscopic identification before and after melting; smallest detected size: 0.49 mm.	MP kg^{-1}	Stainless steel/ glass tools, PP bags for storage, cotton lab coats, and gloves. Unmulched field samples were processed as control, additionally distilled water samples were processed as blank during MP extraction. Triplicate samples analyzed. Method restricted to low-density plastics and did not detect high-density plastics	Meng et al. (2020)

Shouguang City, China.	Agricultural soils	0–5, 5–10, and 10–25 cm	Samples were collected using a stainless-steel shovel		Samples were sieved (5 mm) to remove stones and debris.	Fenton reagent was used for organic matter digestion, under pH 3–5 and <40 °C (ice bath)	200 g soil was mixed with saturated NaCl solution (1.2 g/cm^3), ultrasonicated, and left to settle. Supernatants were collected after 24 hours.	The solutions were centrifuged, and MPs were filtered onto 20-μm glass fiber filters.		A stereo microscope was used for size, shape, and color identification. μ-FT-IR was used for polymer identification.	MP Kg^{-1}	Non-plastic tools, aluminum foil covers, cotton lab coats. Quartz sand was processed as laboratory blanks. Spiked samples were used for soil recovery rates. Triplicate samples were used for recovery estimations.	Yu et al. (2021)
Hangzhou Bay, China.	Farm soil Collected from 15 areas	0–10 cm	Samples were collected using a stainless shovel.			Organic matter was digested with H_2O_2 (30%, at 70 °C)	The density separation was done using NaI, $\rho = 1.6$ g/cm^3)	Dispersed soil in sodium hexameta-phosphate, followed by saturated NaCl solution with continuous airflow to float microplastics and filtered onto 20 μm glass fiber filter.		Stereomicroscope used to identify MPs and Polymer identification using μ-FT-IR.	MP Kg^{-1}	Non-textile lab attire, glassware, and dedicated microplastic lab. Deionized water was analyzed as blank. Spiked samples were used to calculate recoveries. Duplicates were analyzed.	Zhou et al. (2020)

(continued)

Table 3.1 (continued)

Country/ region	Sampling			Sludge Samples		Sample extraction				Identification	Units reported	Quality control	
	Matrix	Sampling depth	Pre-treatment	Matrix	Treatment	Pre-digestion	Flotation	Filtration	Post-oxidation				Reference
Valencia, Spain (11 with sewage sludge applications, 5 without) sampled.	Agricultural soil and sewage sludge	0–10 and 10–30 cm		Collected from 4 wastewater treatment plants	Samples were dried at 40 °C for 72 hours		NaI was used for density solution of heavier density MPs and lighter MPs were seperated using centrifuge with distilled water at 3000 RPM for 10 minutes.	Filtered using 11 μm pore size.		Microscopic examination done pre-/post-heating (120–130 °C). MPs were identified by increased circularity, transparency, and shininess. μFTIR analysis for polymer identification	MP Kg^{-1}	Cotton lab coats, dedicated microplastic lab. Deionized water was analyzed as blank to correct the results. Spiked samples were also analyzed to assess recovery. Duplicates were also analyzed. Organic matter interference noted; no natural blank samples available.	van den Berg et al. (2020)
Guangxi Province, China	Agricultural soils from citrus orchards	0–5, 5–15, and 15–25 cm.		Dewatered sludge from 3 wastewater treatment plants		Digestion with 30% H_2O_2 at 65 °C for 72 hours.	Density separation using $ZnCl_2$ and NaCl solutions.- Ultrasonication (15 min) to remove adhered particles.	Filtration through 25 μm and 0.45 μm filters.		Stereomicroscopic identification based on morphology and μ-FTIR for identification of polymer	MP Kg^{-1}	Cotton lab coats, and nitrile gloves were used. Containers rinsed with ultrapure water; openings covered with aluminum foil. Blank samples were analyzed as well as triplicates.	Zhang et al. (2020)

62.50 ± 12.97 items kg^{-1} in the deeper layer. However, it should be considered that agricultural soils are often tilled, so the top layer is most likely not undisturbed (van den Berg et al. 2020). On the other hand, nonagricultural, coastal, or floodplain soils are usually undisturbed, and the top 5 cm is representative of surface MP contamination in soil (Piehl et al. 2018; Xu et al. 2020).

3.2.3 Soil Sample Handling

A large sample volume of 0.5–2.0 kg should be collected to enhance representativeness and accommodate replicates for moisture, recovery rates, and physicochemical testing (Möller et al. 2020). Post-collection, samples should be stored in non-plastic containers, e.g., glass jars, stainless steel jars, aluminum foil trays, and sealed to prevent contamination (Zhou et al. 2020). Containers must be clean, dry, and labeled with location, depth, date, and collector details, and preferably stored at 4 °C.

3.2.4 Pre-Treatment

The soil samples are to be cleaned by removing roots and large debris, prior to drying and processed to generate standardized moisture-free baseline data for MP in soil. Different research groups have used varied strategies from air-drying at room temperature to oven-drying and freeze-drying. Details are summarized in Table 3.1. Liu et al. (2018) dried the samples at 70 °C for 24 hours, while van den Berg et al. (2020) followed ISO standards and samples were dried at 40 °C for 72 hours. Several studies have highlighted the deterioration of MP due to higher temperatures (Uddin et al. 2020a, b, 2022a, b, 2024; Jinadasa et al. 2023; Lee et al. 2024). Deterioration is reported in polyethylene (PE) and polyamide (PA) just over 50 °C, while 70 °C is generally safe for most other polymers (Hurley et al. 2018). Hence, to minimize structural changes, drying temperatures should remain below 60 °C, preferably ≤40 °C. Freeze-drying, though an alternative, is less practical due to limited throughput (Dorau et al. 2023) and often the clay-rich soils become too hard (stone-like) making the MP extraction cumbersome and impractical.

Sieving is performed by most workers; however, the choice of sieves is often varied. Some researchers have used a cascade of sieves to fractionate different size fractions, while others have used a 2 mm sieve that allows smaller particles to pass and retains those >2 mm. The MPs in the 2–5 mm range can be visually isolated. A major challenge lies in retaining nanoparticles and fine MPs during separation, as their small size increases the risks of loss or misclassification, thereby skewing quantification (He et al. 2018). The different approaches used for sieving have been summarized in Table 3.1. The study from China (Nanjing) and Switzerland used coarse sieving with 1–5 mm sieves to remove debris and MPs. In another study from

Dian Lake Basin, China, a fine sieve of <1 mm was used, retaining MPs and excluding aggregates. In studies reported from Spain and Taiwan, the debris, including roots, stones, and macroplastics, was manually removed. Enhancing recovery requires refined techniques, such as employing finer meshes or specialized filtration membranes optimized for capturing sub-millimeter MPs. These adjustments improve accuracy in detecting and characterizing MPs, offering deeper insights into their prevalence and behavior in soil ecosystems.

3.3 Sample Preparation

The agricultural soil, especially where the sewage sludge, compost, and fertilizers are applied, is organic-rich, and for accurate quantification of MP in soils, organic matter (OM) removal is a crucial step (Uddin et al. 2020a, 2021, 2022a, b, 2024; Lee et al. 2024; Naji et al. 2021). The organic matter removal should precede the density separation. It has been observed that sonication and density separation often are inefficient as MP embedded in organic matrixes or OM adhered to MPs like PET, nylon, etc., have a similar density range (1.0–1.4 g cm^{-3}) (Bläsing and Amelung 2018). It is quite complex to remove OM without compromising the integrity of polymers, as some of the processes are exothermic, and poorly controlled processes (when temperature gets very high) can compromise the integrity of plastic polymers. In the case of agricultural soils, the OM is also enriched with diverse plant and animal residues that have very different properties and react differently. The more reactive fraction of OM includes rapidly decomposable components such as lipids, proteins, and sugars (Bronick and Lal 2005), which can interfere with MP detection and must be eliminated during sample pretreatment. Organic matter removal typically involves chemical treatment with acids, alkalis, oxidizers, or enzymatic digestion (Bläsing and Amelung 2018; Fischer and Scholz-Böttcher 2017; Loder et al. 2017; Steinmetz et al. 2020; Thomas et al. 2020).

The choice of density separation fluid is equally important for the success of MP segregation from the matrix. Most researchers use sodium chloride (NaCl), which has a density of 1.2 g cm^{-3} as it is cost-effective; however, it has limited efficiency for high-density polymers. Also, it is not clear how many researchers have used analytical grade NaCl, since 94.4% of table salt is known to have MPs (Özçifçi et al. 2023). Some studies have used distilled water, sodium bromide (NaBr), calcium chloride ($CaCl_2$), zinc chloride ($ZnCl_2$), zinc bromide (ZnBr) and sodium iodide (NaI) to name a few. The efficacy of MP removal is quite high with NaI compared to NaCl, but its cost is prohibitive, approximately 70 times costlier than NaCl in some countries. While $ZnCl_2$ is not very environmentally friendly, it has been observed that some research groups do a density separation in a single step, e.g., a study from Mauritius using NaCl, while others have done a sequential separation, repeating the density separation several times using $ZnCl_2$ or NaI, e.g., the study from Chile and Ontario (Table 3.1).

Organic matter digestion can be achieved through various chemical methods, including acids, hydrogen peroxide (H_2O_2), and Fenton's reagent. The use of 65% nitric acid (HNO_3) or H_2SO_4 effectively removes OM, achieving <1% residual organics after 24 hours at 90 °C, but it risks degrading polymers like acrylonitrile butadiene styrene (ABS), polyamide (PA), and polyethylene terephthalate (PET). Among the studies reviewed, a Swiss floodplain study used acid digestion.

Various workers have also used alkaline solutions, such as a 1:1 KOH-NaCl mixture, which can mitigate plastic degradation while degrading OM.

In contrast, hydrogen peroxide (H_2O_2) is widely used for OM oxidation, with studies employing concentrations of 30–35% to disperse soil aggregates and digest organics, achieving up to 72% recovery for microplastic (MP) when applied to high-OM soils. For instance, 35% H_2O_2 effectively dispersed 72% of MPs bound to aggregates, though prolonged treatments (e.g., 72 hours at 23 °C) may be required. This is the most commonly used method; several studies reported from China, Chile, and Greece (Table 3.1) have used H_2O_2 with temperatures ranging from 50–70 °C and reaction time varying between 24 and 72 h. This reaction is exothermic, and it is highly recommended that samples be put in an ice bath to lower the temperature and preserve the polymer integrity.

Fenton's reagent, an advanced oxidation method combining H_2O_2 with Fe^{2+} under acidic conditions (pH 3–5), generates hydroxyl radicals that degrade OM with near-complete efficiency (100 ± 10%) at ambient temperatures. This method has shown promise in organic-rich sediments, recovering 90% of polyethylene (PE) without high-temperature inputs, though it remains under evaluation for broader soil MP separation applications. Fenton's reagent provides high degradation efficacy under mild conditions; each method requires context-specific optimization to minimize plastic damage and maximize organic clearance. The studies from Ontario, Canada, and Shouguang, China (Table 3.1) used Fenton's reagent for recalcitrant organics. Further research is needed to standardise these protocols, particularly for Fenton-based approaches.

The filtration of soil samples for microplastic analysis is a critical step that requires careful selection of filter types, pore sizes, and polymer compositions to ensure efficient capture and accurate identification. Membrane filters, commonly fabricated from polymers such as polycarbonate, nitrocellulose (or mixed cellulose esters), polytetrafluoroethylene (PTFE), and nylon, are widely employed due to their distinct chemical and physical properties. Polycarbonate membranes are favored for their smooth, uniform surface and high chemical resistance, facilitating clear microscopic imaging and Fourier-transform infrared (FTIR) or Raman spectroscopy without spectral interference. Nitrocellulose filters, while cost-effective and suitable for routine use, may require caution due to potential biodegradability in harsh chemical treatments. PTFE membranes, known for their hydrophobicity and exceptional resistance to aggressive reagents, are ideal for organic solvent-based digestion processes. Pore sizes typically range from 0.45 μm to 5 μm, with 0.45 μm being standard for capturing small microplastics (≥1 μm), though larger pores (e.g., 1.2–5 μm) may be used to prevent clogging from soil particulates, sometimes in tandem with pre-filtration steps. The choice of pore size balances the need to retain target microplastics while managing sample load and filter longevity. The choice of pore size should also take into account the techniques to be used for identification, i.e., if an ATR-FTIR is to be used, MPs <300 μm are usually not attempted, while if micro-Raman spectroscopy is used, technically one can identify polymer up to 1 μm size.

Polymer selection also considers compatibility with downstream analytical techniques; for instance, nylon's hydrophilic nature suits aqueous solutions, whereas PTFE's inertness avoids spectral overlap with common synthetic polymers like polyethylene (PE) or polypropylene (PP). If Nile Red solution is to be used for staining the MPs, a polymer filter is to be avoided. Glass fiber filters, though non-polymeric, are occasionally used for high-throughput sieving but lack the precision of polymer membranes. Among the studies reviewed, the filter type is documented whenever provided in Table 3.1. In the study from Switzerland cellulose acetate filter with 0.45 μm pore size was used, which is prone to degradation during processing. The study indicated that nanoplastics were captured, but the risk of clogging was high. On the other hand, the studies from China have used nylon membrane filters and glass fiber filters in a wide range of pore sizes from 0.45 to 25 μm. High-quality nylon filters of 20 μm were shown to be efficient in a study from Shaanxi, China, whereas in another study from Xinjiang, China, the large-pore filters (>50 μm) were reported to have probably underestimated small MPs. Ultimately, the interplay of filter material, pore size, and soil matrix characteristics ensures optimal recovery of microplastics while minimising contamination and analytical interference, crucial for reliable environmental monitoring.

3.4 Microplastic Identification

The identification of MPs is usually done in two steps: a microscopic examination followed by a polymeric characterization using one of the chemical characterization approaches.

3.4.1 Microscopic Characterization

The microscopic identification of MPs is the most prevalent approach using a compound, dissection, stereomicroscope, or UV microscope. It provides a first overview of the likely MP's presence in a sample based on visual inspection and physical characteristics (Uddin et al. 2020a, 2022a). The absence of cellular structure, consistency in thickness along the length of the particle, consistent colour, no bending or tapering, clarity, and higher relief to name a few are characteristics of a polymer particle. Based on the microscopic examinations, the shape of possible MPs is also identified as fibre, fragment, film, and sphere. Several researchers have highlighted the use of fluorescent staining dyes such as Nile Red (Hidalgo-Ruz et al. 2012) to improve the detection of MPs under a UV microscope. However, it has been shown by subsequent workers that the staining is not exclusive to MPs only; a lot of natural materials like wood, wool, chitin, and natural fibres also are stained, leading to significant misidentification (20–70%) of smaller MPs <100 μm (Song et al. 2015). In the reviewed studies (Table 3.1) most of them have used microscopic examination. Some have done morphotyping of MPs into fibres, fragments, and films, like the study from Shanghai, China. In another study from Murcia, Spain, the MP size determination was done using ImageJ software. Despite its accessibility and low cost, microscopic identification alone is considered insufficient for definitive MP identification and has resulted in false positives where organic fibers have frequently been misidentified as MPs.

3.4.2 *Polymeric Characterization of Microplastics*

To overcome the issue of false positives, it was realised that polymer validation is crucial (Shim et al. 2017). The most commonly applied techniques for polymer validation are Fourier-transform infrared (FTIR) and micro-Raman spectroscopy (Mariano et al. 2021). Additionally, automated digital imaging systems and machine learning algorithms are increasingly integrated to reduce human bias and enhance throughput (Bianco et al. 2020; Beres et al. 2023).

3.4.2.1 Fourier-Transform Infrared (FTIR) Spectroscopy

The Fourier-transform infrared spectroscopy identifies polymers by detecting bond specific infrared absorption, differentiating MPs from organic or inorganic contaminants (Loder et al. 2017). The common operational modes in FTIR include reflectance and transmission, where the transmission mode requires thin, transparent samples, whereas the reflectance mode can work with irregularly shaped and opaque particles (Fok et al. 2020). In the study from the Netherlands, the transmission mode was used, and the standard particle size was >500 μm (Table 3.1). The most commonly used FTIR for microplastic analyses is attenuated total reflectance-FTIR (ATR-FTIR) which is quite effective for larger particle sizes (≥300 μm) and was used by the study conducted in Greece (Table 3.1). In the ATR-FTIR, we have to put the sample manually on the holder, and it is quite cumbersome if a large number of MPs are present in the sample. The micro-FTIR is coupled with microscopy, and it can effectively identify particles as small as 10 μm. Current models of micro-FTIR often allow semi-automated mapping and can streamline workflows and improve throughput. The data generated are often referenced against spectral libraries, and different workers use different thresholds; the most commonly accepted threshold is ≥70% match.

3.4.2.2 Raman Spectroscopy

The use of micro-Raman spectroscopy enables identifying MPs down to 1 μm with micro-Raman systems (Kappler et al. 2016). The underlying principle behind the instrument leverages the Raman effect—frequency shifts in scattered light—to identify polymers by analyzing vibrational modes and molecular structures (Smith and Dent 2019). It generates unique spectral fingerprints non-destructively in most cases, but there are reports of MP burning from prolonged laser exposure. A study reported in Table 3.1 from the Tibetan Plateau used micro-Raman and detected pigments and additives as well. The technique struggles with fluorescent additives or pigments, necessitating rigorous sample purification. However, the fluorescence interference is often mitigated by applying baseline correction algorithms and using advanced detectors. Despite its strengths—high specificity, spatial resolution, and non-destructive nature—Raman spectroscopy faces limitations such as weak signal intensity, lengthy data processing, and complex sample preparation. These micro-Raman systems are more costly compared to an ATR-FTIR and micro-FTIR.

3.4.2.3 Pyrolysis-Gas Chromatography Mass Spectrometry (Py-GC/MS)

Pyrolysis-gas chromatography-mass spectrometry (Py-GC/MS) has been used by some research groups. The Py-GC/MS thermally decomposes MPs under oxygen-limited conditions, identifying polymers via their degradation products (Hermabessiere et al. 2018). Double-shot Py-GC/MS enhances efficiency by performing sequential thermal desorption at varying temperatures. Combined with pressurized liquid extraction (PLE), it achieves low detection limits (14 $\mu g\ g^{-1}$ for tyre particles) and effectively quantifies PE, PP, PMMA, and PS (Okoffo et al. 2020). Another novel technique recently developed is the pyrolysis-gas chromatography coupled to Orbitrap-mass spectrometry (Py-GC/OMS) (Torres-Agullo et al. 2024). The Py-GC/OMS has been optimized for several polymers including polymethyl methacrylate (PMMA), cyclopentanone (nylon-66), polypropylene (PP), nitrile butadiene rubber (NBR), polyvinyl chloride (PVC), polyethylene terephthalate (PET), acrylonitrile butadiene styrene (ABS), polyethylene (PE), polycarbonate (PC), and polystyrene (PS) (Torres-Agullo et al. 2024). The use of Py-GC/MS allows the generation of data as polymer mass in the matrix and is often incomparable with the concentration data reported in number per unit mass or volume. The other issue is the technique doesn't provide insight into the size of MP or nanoplastic, which is again quite crucial for toxicity assessment. However, we have not seen any instance of its usage for MP determination in soil.

3.4.2.4 Laser Direct Infrared Spectroscopy (LDIR)

The emergence of LDIR as a transformative tool for microplastic research in the past few years is not only based on its ease of use and faster processing time, but also on its technological superiority. The LDIR utilizes a high-resolution camera that is capable of resolving particles up to 10 μm and, coupled with the quantum cascade laser (QCL), its response time is quick and almost instantaneous. It acquires the entire infra-red spectrum for an MP in just one second (Isaac Chandran and Veerasingam 2025). The technique provides an excessive advantage in terms of time and sample preparation compared to any other conventional technique (FTIR, Micro-Raman or Py-GC/MS), making it very valuable for matrixes where there is a large number of MPs. Within the studies reviewed only the study from Shandong, China, has used LDIR for 20–1000 μm MPs.

3.4.2.5 Thermal Analysis

Thermal analyses such as thermogravimetric analysis (TGA) and differential scanning calorimetry (DSC) are destructive techniques that are used for MP characterization (Isaac Chandran and Veerasingam 2025). These techniques chemically characterize MPs by assessing thermal degradation profiles (Dümichen et al. 2017). The efficacy of the thermal analysis is limited in differentiating between similar polymers, like PVC, polyurethane, PET, and polyesters. Despite these constraints, DSC remains advantageous for its simplicity and rapid results.

3.4.2.6 Scanning Electron Microscopy (SEM)

The scanning electron microscope (SEM) paired with the energy-dispersive X-ray spectroscopy (EDS) is an imaging technique that analyzes MPs' surface morphology and elemental composition (Sadri and Thompson 2014). However, its application is limited by high costs, time-consuming protocols, and challenges in identifying non-distinct particles (Primpke et al. 2020) making it less suitable for routine monitoring. In a study reported within this review, a study from Tamil Nadu, India, used SEM-EDS for studying the surface morphology and elemental composition of MPs. The complex sample preparation, including drying, and conductive coating the sample, can obliterate the MP morphology as well. It is a destructive technique with limited capability of polymer identification.

3.4.2.7 Liquid Chromatography-Mass Spectrometry (LCMS)

Liquid chromatography-mass spectrometry (LC-MS) has emerged as a powerful analytical tool for identifying and quantifying microplastics, sub-micron particles, or chemically degraded polymers. The technique leverages chromatographic separation to isolate analytes, followed by mass detection, often employing electrospray ionization (ESI) for polar polymers or additives like phthalates. However, LC-MS enables targeted analysis of polymer-specific degradation products, additives, or monomers, offering high sensitivity and specificity in complex environmental matrices. While LC-MS excels in detecting dissolved or chemically extracted components, its application requires rigorous sample preparation, including enzymatic or oxidative digestion to remove organic interference without altering target polymers. It is prudent to highlight the challenges in universal polymer applicability and its excellent ability to detect PET and PS; LC-MS complements existing methods by providing quantitative insights into subvisible particles and additive leaching, advancing environmental monitoring and toxicological studies of microplastics. The information provided will not include size and shape data and will be comparable to information derived from Py-GCMS.

3.4.2.8 Atomic Force Microscopy (AFM)

The AFM generates nanometer-resolution images by scanning samples in contact or non-contact modes. When integrated with Raman or IR spectroscopy, it provides simultaneous topographic and chemical data. For instance, AFM-IR detects thermal expansion from IR absorption, translating it into frequency patterns via Fourier analysis that can be used for MP identification (Dazzi et al. 2017). However, it is not very commonly used until now.

3.5 Quantification of Microplastics and Reporting Units

The quantification of MPs in the absence of standardised methodologies and reporting units has introduced significant challenges in data comparability and interpretation. Researchers employ diverse analytical techniques and units of measurement, leading to the generation of incomparable datasets, making any spatio-temporal assessments challenging and complicated. This issue is particularly pronounced in soil and agricultural systems, where heterogeneous matrices amplify methodological disparities. Despite these issues, most researchers, about 80% of the studies reviewed have reported MP abundance in terms of MP Kg^{-1}. Among the studies reviewed in this paper, a study from Taiwan reported MP abundance in surface soil as MP m^{-2}. In some studies, like from Xinjiang, China, where mulch was added to soil, the MPs are reported as g m^{-2}. The results from studies where Py-GC/MS, LCMS, and TGA are used results are often reported as mg Kg^{-1}, though they are not frequently used for MP in soils.

3.6 Quality Assurance and Quality Control

Robust quality assurance and quality control (QA/QC) protocols are essential to ensure the integrity of microplastic (MP) analyses across all stages, from field sampling to laboratory processing and identification. Contamination risks—arising from airborne MPs, laboratory equipment, or reagents—must be systematically minimized to safeguard the precision and validity of results. To address these challenges, stringent measures should be implemented, including the use of metal tools and avoiding plastic materials. Most of the studies reviewed in this paper have avoided using plastic tools. Several workers have reported thorough cleaning of instruments with Milli-Q water and used cotton lab coats. Most workers have covered the samples with aluminium foil to minimize airborne contamination during sample preparation, and few have used laminar flow cabinets and maintained clean-air environments, supported by high-efficiency particulate air (HEPA) filtration systems to reduce airborne contamination (Schell et al. 2022; Möller et al. 2020). Procedural blanks, including container and solution blanks, are processed identically to samples to monitor background contamination. About 90% of the studies reviewed in this communication have used blanks. Additionally, recovery tests using spiked samples have been used by many research groups. High recoveries of PE/PP (80–95%) have been reported by workers using $NaCl/ZnCl_2$, while low recovery was reported for PET/PVC, essentially as they used NaCl and these polymers are denser than the separating fluid; one such report was from Shanghai, China. Another important aspect of QA/QC is the replicate samples. The study from Germany and Spain analyzed triplicate samples, whereas Tanzania and Mexico analysed a single sample. The QA/QC strategies are to be strictly followed to enhance the reliability of MP inventories.

3.7 Methodological Challenges and Recommendations

There have been significant differences among studies in sample collection, preparation, and identification. These differences are evident since the soil is a heterogeneous matrix, and also because there is a huge disparity in the research resources and scientific questions addressed.

Based on the studies reviewed, it is observed that studies need to adopt standardized protocols; some of the key considerations to be taken are summarized below.

3.7.1 Sampling Strategies

Since soil is a heterogeneous matrix, collection of representative samples is a key to generating realistic data. It is highly recommended to undertake composite sampling by combining multiple subsamples across the study area. In agricultural systems, where there is high plastic input due to the use of mulch, sludge application, compost, or use of wastewater for irrigation, due consideration should be given not to skew the sampling to the topsoil only. In order to capture the true extent of MP pollution in agrosystems, it is recommended that stratified samples from 0–10 cm, 10–20 cm, 20–30 cm, and 30–40 cm be collected to assess vertical distribution, especially in tilled fields. Depending on the size of the study area, the sampling can either be grid or random. It is extremely important to note down the GPS coordinates and all relevant information on agricultural practices like tillage, crops grown, irrigation practices, and whether soil additives, sludge, mulch, etc., have been used.

3.7.2 Extraction and Separation

From the published papers it is explicit that most studies have done a density separation using distilled water, NaCl, NaI, sodium polytungstate (NaPT), or zinc chloride ($ZnCl_2$). The recovery rates with distilled water are expected to be very low due to the negative buoyancy of most polymers. However, density separation is very high with $ZnCl_2$ but it is corrosive for high recovery rates. We recommend NaI though it is much more expensive than NaCl. It has also been observed that centrifugation enhances MP removal efficiency. Additionally, sonication has proven to be beneficial, but it should be low-energy sonification else MPs can also disintegrate into smaller sizes. For the digestion of organic matter from soil, 30% H_2O_2 is quite effective. However, KOH or enzymatic digestion with cellulase or protease is effective and would minimize polymer degradation. For clay-rich soils, adding a dispersant like sodium hexametaphosphate during extraction improves the extraction of MPs.

3.7.3 Identification and Quantification

Most MP research groups use a multi-threaded approach since the mere microscopic examination is not sufficient for the identification of tiny plastic particles. The first step toward the MP identification is microscopic examination using a stereo-microscope, compound microscope, dissection microscope, or UV microscope to identify the morphology (size, shape, colour). However, for polymer identification, one of the several analytical techniques discussed in this paper can be used; the most commonly used polymer identification techniques are ATR-FTIR (>300 μm), micro-FTIR (20 μm), and micro-Raman spectroscopy (1 μm).

Some advanced techniques like Py-GC/MS, Py-GC/OMS, and LC-MS can provide mass quantification of MPs and even for nanoplastics. However, they have limited usage in MP quantification in soil; it might have high usage if food chain transfer from soil to edible plants is to be looked at.

3.7.4 Data Reporting

The data is usually reported as particle count per unit mass of matrix, which is the soil in this case; thus, the results are to be reported as MP Kg^{-1}. Some research groups also report mass concentrations as mg Kg^{-1}, but that is only feasible in soils where high MP concentration is seen (i.e., sludge applied or mulched soils). It is important to report the size range of MPs as well.

In the case of soil, the metadata is also important, as details of soil texture, grain size, organic matter content, agricultural inputs (fertilizers, irrigation), and coordinates should be recorded.

3.7.5 Quality Control

Looking at the wide range of methodologies and practices employed for MP determination in soils, the QA–QC becomes more so important to generate a reliable dataset. Use of cotton lab coats, glass and metal containers, and working in a laminar flow unit is highly recommended. Also, the use of both field and lab blanks will provide information for realistic correction of data. Some of the studies have used spiked samples to validate methods, and it is also highly recommended.

3.8 Conclusion

This summary of studies so far on microplastics in soil highlights the limited work carried out so far, even though gigantic amounts of MPs end up in agricultural lands as a result of the use of sewage, compost for amending low-productivity soils, use of wastewater for irrigation, and mulches for improved productivity. Only 39 studies were published where details of the methodology are reported for MPs in soil. A quick overview highlights that the methods used for soil are modified methods for MP determination in marine and freshwater matrixes. One very conspicuous difference between soil and other matrixes is the heterogeneity of soil, which makes it imperative that representative samples are collected. The presence of fine-grained clastics, mainly clays, warrants specific steps to ensure the segregation of MPs from the matrix. The digestion of organic matter using 30% H_2O_2 or enzymatic digestion seems quite effective. The use of Py-GC/MS and LCMS for MPs in soil gives a very different level of information on the mass of polymers rather than number. Though not very popular, we have found LDIR to be very effective in the detection of MPs; it provides both the size information and numbers and type of polymers with reasonably high accuracy. There is certainly a need to harmonize methods to be used for MPs in soil for generating comparable datasets across spatial domains. Future work should address nanoplastics and biodegradable polymers, ensuring methodologies evolve with agricultural plastic use.

References

Bello FA et al (2025) Microplastics in agricultural soils: sources, impacts on soil organisms, plants, and humans. Environ Monit Assess 197(4):448

Beres ND et al (2023) Merging holography, fluorescence, and machine learning for in situ, continuous characterization and classification of airborne microplastics. EGUsphere 2023:1–31

Beriot N et al (2021) Low density-microplastics detected in sheep faeces and soil: a case study from the intensive vegetable farming in Southeast Spain. Sci Total Environ 755:142653

Bianco V et al (2020) Microplastic identification via holographic imaging and machine learning. Adv Intell Syst 2(2):1900153

Bigalke M et al (2022) Microplastics in agricultural drainage water: a link between terrestrial and aquatic microplastic pollution. Sci Total Environ 806:150709

Bläsing M, Amelung W (2018) Plastics in soil: analytical methods and possible sources. Sci Total Environ 612:422–435

Boughattas I et al (2021) Assessing the presence of microplastic particles in Tunisian agriculture soils and their potential toxicity effects using Eisenia andrei as bioindicator. Sci Total Environ 796:148959

Bronick CJ, Lal R (2005) Soil structure and management: a review. Geoderma 124(1):3–22

Choi Y et al (2021) Plastic contamination of forest, urban, and agricultural soils: a case study of Yeoju City in the Republic of Korea. J Soils Sediments 21:1–12

Corradini F et al (2019) Evidence of microplastic accumulation in agricultural soils from sewage sludge disposal. Sci Total Environ 671:411–420

Crossman J et al (2020) Transfer and transport of microplastics from biosolids to agricultural soils and the wider environment. Sci Total Environ 724:138334

Dazzi A and CB, Prater 2017 AFM-IR: Technology and applications in nanoscale infrared spectroscopy and chemical imaging. Chemical reviews, 117(7): p. 5146–5173Ding L et al (2020) The occurrence and distribution characteristics of microplastics in the agricultural soils of Shaanxi Province, in north-western China. Sci Total Environ 720:137525

Dorau K et al (2023) Status quo of operation procedures for soil sampling to analyze microplastics. Microplast Nanoplast 3(1):15

Dümichen E et al (2017) Fast identification of microplastics in complex environmental samples by a thermal degradation method. Chemosphere 174:572–584

Fakour H et al (2021) Quantification and analysis of microplastics in farmland soils: characterization, sources, and pathways. Agriculture 11:330

Feng S et al (2020) Analysis of microplastics in a remote region of the Tibetan Plateau: implications for natural environmental response to human activities. Sci Total Environ 739:140087

Fischer M, Scholz-Böttcher BM (2017) Simultaneous trace identification and quantification of common types of microplastics in environmental samples by pyrolysis-gas chromatography–mass spectrometry. Environ Sci Technol 51(9):5052–5060

Fok L et al (2020) A meta-analysis of methodologies adopted by microplastic studies in China. Sci Total Environ 718:135371

Harms IK et al (2021) Amount, distribution and composition of large microplastics in typical agricultural soils in Northern Germany. Sci Total Environ 758:143615

Hartmann NB et al (2019) Are we speaking the same language? Recommendations for a definition and categorization framework for plastic debris. Environ Sci Technol 53(3):1039–1047

He D et al (2018) Microplastics in soils: analytical methods, pollution characteristics and ecological risks. TrAC Trends Anal Chem 109:163–172

He D, Zhang X, Hu J (2021) Methods for separating microplastics from complex solid matrices: comparative analysis. J Hazard Mater 409:124640

Hermabessiere L et al (2018) Optimization, performance, and application of a pyrolysis-GC/MS method for the identification of microplastics. Anal Bioanal Chem 410(25):6663–6676

Hernández-Gutiérrez E et al (2021) Morphospecies abundance of above-ground invertebrates in agricultural systems under glyphosate and microplastics in South-Eastern Mexico. Environments 8(11):130
Hidalgo-Ruz V et al (2012) Microplastics in the marine environment: a review of the methods used for identification and quantification. Environ Sci Technol 46(6):3060–3075
Huang Y et al (2020) Agricultural plastic mulching as a source of microplastics in the terrestrial environment. Environ Pollut 260:114096
Hurley RR et al (2018) Validation of a method for extracting microplastics from complex, organic-rich, environmental matrices. Environ Sci Technol 52(13):7409–7417
Isaac Chandran PJ, Veerasingam S (2025) Laser direct infrared spectroscopy: a cutting-edge approach to microplastic detection in environmental samples. Talanta 284:127284
Isari EA et al (2021) Microplastics in agricultural soils: a case study in cultivation of watermelons and canning tomatoes. Water 13(16):2168
ISO (2017) ISO 18400-102 soil quality — sampling — part 102: selection and application of sampling technique. In: I.S. Organization (ed) First edition 2017–01, Switzerland
Jinadasa BKKK, Uddin S, Fowler SW (2023) Microplastics (MPs) in marine food chains: is it a food safety issue? In: Ozogul F (ed) Advances in food and nutrition research. Academic Press, pp 101–140
Kappler A et al (2016) Analysis of environmental microplastics by vibrational microspectroscopy: FTIR, Raman or both? Anal Bioanal Chem 408:8377
Katsumi N et al (2021) Accumulation of microcapsules derived from coated fertilizer in paddy fields. Chemosphere 267:129185
Kawecki D, Goldberg L, Nowack B (2021) Material flow analysis of plastic in organic waste in Switzerland. Soil Use Manag 37(2):277–288
Kim S-K et al (2021) Abundance and characteristics of microplastics in soils with different agricultural practices: importance of sources with internal origin and environmental fate. J Hazard Mater 403:123997
Kumar MV, Sheela AM (2021) Effect of plastic film mulching on the distribution of plastic residues in agricultural fields. Chemosphere 273:128590
Kumar M et al (2020) Microplastics as pollutants in agricultural soils. Environ Pollut 265:114980
Kundu MN, Komakech HC, Lugomela G (2022) Analysis of macro- and microplastics in riverine, riverbanks, and irrigated farms in Arusha, Tanzania. Arch Environ Contam Toxicol 82(1):142–157
Lee J-Y et al (2024) A comprehensive review of urban microplastic pollution sources, environment and human health impacts, and regulatory efforts. Sci Total Environ 946:174297
Li Q et al (2019) Separation and identification of microplastics from soil and sewage sludge. Environ Pollut 254:113076
Li W et al (2020) Microplastics in agricultural soils: extraction and characterization after different periods of polythene film mulching in an arid region. Sci Total Environ 749:141420
Li Q et al (2021) Are microplastics correlated to phthalates in facility agriculture soil? J Hazard Mater 412:125164
Li S et al (2022) Macro- and microplastic accumulation in soil after 32 years of plastic film mulching. Environ Pollut 300:118945
Liu M et al (2018) Microplastic and mesoplastic pollution in farmland soils in suburbs of Shanghai, China. Environ Pollut 242:855–862
Loder MGJ et al (2017) Enzymatic purification of microplastics in environmental samples. Environ Sci Technol 51(24):14283–14292
Mariano S et al (2021) Micro and Nanoplastics identification: classic methods and innovative detection techniques. Front Toxicol 3:636640
Meng F et al (2020) Effects of plastic mulching on the accumulation and distribution of macro and micro plastics in soils of two farming systems in Northwest China. PeerJ 8:e10375
Möller JN, Löder MGJ, Laforsch C (2020) Finding microplastics in soils: a review of analytical methods. Environ Sci Technol 54(4):2078–2090

Naji A et al (2021) Microplastics in wastewater outlets of Bandar Abbas city (Iran): a potential point source of microplastics into the Persian Gulf. Chemosphere 262:128039

Okoffo ED et al (2020) Identification and quantification of selected plastics in biosolids by pressurized liquid extraction combined with double-shot pyrolysis gas chromatography-mass spectrometry. Sci Total Environ 715:136924

Özçifçi Z, Basaran B, Akçay HT (2023) Microplastic contamination and risk assessment in table salts: Turkey. Food Chem Toxicol 175:113698

Piehl S et al (2018) Identification and quantification of macro- and microplastics on an agricultural farmland. Sci Rep 8(1):17950

Primpke S et al (2020) Critical assessment of analytical methods for the harmonized and cost-efficient analysis of microplastics. Appl Spectrosc 74(9):1012–1047

Ragoobur D, Huerta-Lwanga E, Somaroo GD (2021) Microplastics in agricultural soils, wastewater effluents and sewage sludge in Mauritius. Sci Total Environ 798:149326

Sa'adu I, Farsang A (2022) Greenhouse farming as a source of macroplastic and microplastics contamination in agricultural soils: a case study from Southeast-Hungary. Agrokém Talajt 71(1):43–57

Sadri SS, Thompson RC (2014) On the quantity and composition of floating plastic debris entering and leaving the Tamar Estuary, Southwest England. Mar Pollut Bull 81(1):55–60

Schell T et al (2022) Fate of microplastics in agricultural soils amended with sewage sludge: is surface water runoff a relevant environmental pathway? Environ Pollut 293:118520

Scheurer M, Bigalke M (2018) Microplastics in Swiss floodplain soils. Environ Sci Technol 52(6):3591–3598

Shim WJ, Hong SH, Eo SE (2017) Identification methods in microplastic analysis: a review. Anal Methods 9(9):1384–1391

Smith E, Dent G (2019) Introduction, basic theory and principles. In: Smith E, Dent G (eds) Modern Raman spectroscopy. Wiley, pp 1–20

Song YK et al (2015) A comparison of microscopic and spectroscopic identification methods for analysis of microplastics in environmental samples. Mar Pollut Bull 93(1–2):202–209

Steinmetz Z et al (2020) A simple method for the selective quantification of polyethylene, polypropylene, and polystyrene plastic debris in soil by pyrolysis-gas chromatography/mass spectrometry. J Anal Appl Pyrolysis 147:104803

Tagg AS et al (2022) Agricultural application of microplastic-rich sewage sludge leads to further uncontrolled contamination. Sci Total Environ 806:150611

Thomas D et al (2020) Sample preparation techniques for the analysis of microplastics in soil—a review. Sustainability 12(21):9074

Torres-Agullo A, Zuri G, Lacorte S (2024) Pyr-GC-Orbitrap-MS method for the target/untargeted analysis of microplastics in air. J Hazard Mater 469:133981

Uddin S et al (2020a) Standardized protocols for microplastics determinations in environmental samples from the Gulf and marginal seas. Mar Pollut Bull 158:111374

Uddin S, Fowler SW, Saeed T (2020b) Microplastic particles in the Persian/Arabian Gulf—a review on sampling and identification. Mar Pollut Bull 154:111100

Uddin S et al (2021) A review of microplastic distribution in sediment profiles. Mar Pollut Bull 163:111973

Uddin S et al (2022a) Micro-nano plastic in the aquatic environment: methodological problems and challenges. Animals 12(3):297

Uddin S et al (2022b) Microplastics in Kuwait's wastewater streams. Sustainability 14(23):15817

Uddin AS, Uddin S, Fowler SW (2023) Environmental microplastics: a significant pollutant of the Anthropocene. In: Khan A, Wang C, Asiri AM (eds) Microplastic sources, fate and solution. Springer Nature Singapore, Singapore, pp 89–105

Uddin S et al (2024) Sewage sludge as soil amendment in Arid soils - a trace metal, nutrient and trace organics perspective. Emerg Contamin 11(1):100420

UNEP (2025) Intergovernmental negotiating committee to develop an international legally binding instrument on plastic pollution, including in the marine environment. UNited Nations Environment Program: Fifth session Busan, Republic of Korea, 25 November–1 December 2024 van den Berg P et al (2020) Sewage sludge application as a vehicle for microplastics in eastern Spanish agricultural soils. Environ Pollut 261:114198

van Schothorst B et al (2021) Sources of light density microplastic related to two agricultural practices: the use of compost and plastic Mulch. Environments 8(4):36

Weber A et al (2021) Ingestion and toxicity of microplastics in the freshwater gastropod Lymnaea stagnalis: no microplastic-induced effects alone or in combination with copper. Chemosphere 263:128040

Weber CJ, Santowski A, Chifflard P (2022) Investigating the dispersal of macro- and microplastics on agricultural fields 30 years after sewage sludge application. Sci Rep 12(1):6401

Withana PA et al (2024) Soil microplastic analysis: a harmonized methodology. Crit Rev Environ Sci Technol 54(15):1138–1163

Xu B et al (2020) Microplastics in the soil environment: occurrence, risks, interactions and fate—a review. Crit Rev Environ Sci Technol 50(21):2175–2222

Yang M et al (2021a) Influences of different source microplastics with different particle sizes and application rates on soil properties and growth of Chinese cabbage (Brassica chinensis L.). Ecotoxicol Environ Saf 222:112480

Yang J et al (2021b) Abundance and morphology of microplastics in an agricultural soil following long-term repeated application of pig manure. Environ Pollut 272:116028

Yu L et al (2021) Distribution characteristics of microplastics in agricultural soils from the largest vegetable production base in China. Sci Total Environ 756:143860

Zaman M, Uddin S (2024) Sewage sludge usage in agriculture: are we introducing microplastics in food chain? Soils Newslett, International Atomic Energy Agency 10(2):7–8

Zhang GS, Liu YF (2018) The distribution of microplastics in soil aggregate fractions in southwestern China. Sci Total Environ 642:12–20

Zhang D et al (2016) The status and distribution characteristics of residual mulching film in Xinjiang, China. J Integr Agric 15(11):2639–2646

Zhang L et al (2020) An overlooked entry pathway of microplastics into agricultural soils from application of sludge-based fertilizers. Environ Sci Technol 54(7):4248–4255

Zhang Q et al (2021) Effect of polystyrene microplastics on rice seed germination and antioxidant enzyme activity. Toxics 9(8):179

Zhou B et al (2020) Microplastics in agricultural soils on the coastal plain of Hangzhou Bay, east China: multiple sources other than plastic mulching film. J Hazard Mater 388:121814

Chapter 4
A Simplified and Optimised Protocol for Microplastics Determination in Soil and Sludge Applied Soils

Saif Uddin, Nazima Habibi, Mohammad Zaman, Jin-Yong Lee, and Montaha Behbehani

Abstract Microplastics are emerging contaminants of concern often associated with high concentrations in waste streams. An enormous quantity of dry sludge is generated globally and is often used as a soil amendment in agricultural areas in both the developed and developing countries. The concentration of microplastics (MPs) in sewage sludge varies globally in a wide range between 1 and 186.7 ± 26.0 MP g^{-1}. The methodologies used for MP determination in sludge and soils vary, making any realistic comparison infructuous. The International Atomic Energy Agency (IAEA) took the initiative to develop a simple and optimised method under its Coordinated Research Project (CRP; D15021) to determine MPs in agricultural soils and intends to evaluate the fate and environmental impact of plastics in soil and crops. The methodology for MP determination in agricultural soils and sludge-applied soils is presented. Details are given on sample collection and the need for composite samples for non-homogeneous matrices like soil, sample processing including a preferred method for oxidation of organic matter, density separation fluids and multi-tier sample identification techniques. Adaptation of the method will not only lead to the generation of comparable data for the CRP participants. It will also assist the 178 IAEA member states across the regions in generating analogous datasets on microplastics (MPs) in soil and sludge-applied soils. The data and information generated on MP inventories can be used to underpin the policy on safe sludge usage in agricultural soils.

S. Uddin (✉) · N. Habibi · M. Behbehani
Environment and Life Sciences Research Center, Kuwait Institute for Scientific Research, Safat, Kuwait
e-mail: sdin@kisr.edu.kw

M. Zaman
Joint FAO-IAEA Centre of Nuclear Techniques in Food and Agriculture, Department of Nuclear Sciences and Applications, International Atomic Energy Agency, Vienna, Austria

J.-Y. Lee
Department of Geology, Kangwon National University, Chuncheon, Republic of Korea

S. Uddin, M. Zaman (eds.), *Micro- and Nano-plastics in Soil and Crop Systems*, Sustainability Sciences in Asia and Africa, https://doi.org/10.1007/978-981-95-2740-3_4

4.1 Introduction

4.1.1 Background

Plastics have become an integral part of modern society, offering a wide array of desirable properties such as lightweight, chemical resistance, durability, and affordability. The impact of microplastics (MPs) in terrestrial ecosystems, particularly in soils and soils amended with sludge, has been relatively overlooked. The primary sources of MPs in terrestrial ecosystems were observed with the combination of wastewater sludge and municipal solid waste composts, considered resourceful components for land application as a substitute for traditional fertilisers (Zere Taskin and Bilgili 2023). Sewage sludge from wastewater treatment plants emerges as a major contributor to MP pollution in soil, as it contains both primary and secondary MPs. Despite its potential as a valuable soil additive due to its organic and nutrient content, concerns have arisen regarding the spread of pollutants through sludge application in agriculture.

The MPs have the potential to be transported by the wind and as rain overflows to nearby lands and aquatic systems, besides the vertical movement in the soil profiles due to infiltration of irrigation water and rainwater. Globally, approximately 45 million tons of dry sludge are produced annually (Gao et al. 2020) from wastewater treatment plants, with a significant portion being used in landfills, incinerated, or applied to soil as a fertiliser. This practice has raised environmental concerns, particularly regarding the fate and impact of MPs in soil and crops. These MPs can potentially cause phytotoxic effects and can be directly and indirectly linked to human health. Therefore, quantifying them in the soils and sludge is novel information for managing soils and the environmental quality of respective regions.

There is a need to develop a harmonised method for identifying MPs in soil and sludge-applied soils to generate reliable data for plastic pollution. Microplastics can persist in soils for a long time, with estimates ranging from 15 to 34 years of residence time (Zubris and Richards 2005; Weber et al. 2022). A recent study highlights the uncontrolled contamination of adjacent land due to sludge application, with 44% of the MP load on sludge-applied land found on nearby untreated land, suggesting spillover effects reaching depths of 60–90 cm in the soil profile (Zaman and Uddin 2024). These are important considerations when collecting soil samples for MP assessments.

The lack of standardised protocols for assessing MPs in soil and sludge-applied soil has hindered reliable data collection and comparison across studies. Recognising this need, the Joint FAO-IAEA Programme launched a Coordinated Research project (D15021) to evaluate the fate and environmental impact of plastics in soil and crops. As part of the International Atomic Energy Agency's (IAEA) initiative, a method is developed that deals with the issues emanating from sample collection, sample processing, and techniques used in identification and reporting units in soils and sludge-applied soils. The scientifically sound data would facilitate informed decision-making and design of mitigation strategies to combat MP pollution in terrestrial environments.

4.1.2 *Objective of the Protocol*

The protocol establishes a simplified and optimised method for collecting, processing, and reporting MP abundance in soil samples.

4.1.3 *Scope of the Operational Protocol*

This procedure is applicable for sampling, processing, and determining the abundance of MP particles (1 μm to 5 mm) with densities <1.8 g cm^{-3} in the top 10–20 cm of a soil profile.

4.1.4 *Decision Tree for Selecting the Approach*

We propose to use a simple decision tree approach for MP analyses. This decision tree is based on the type of instrumentation available and the research question to be addressed (Fig. 4.1):

The decision tree depends on the various types of MPs.

Start: Define the scope of MP analysis.

Question 1: What is the size range of interest?

- >5 mm (macroplastics)
  → Proceed with visual identification and manual sorting.
  → Question: Is this size fraction sufficient for the study? If yes, analyse; if no, proceed to smaller sizes.
- 1 mm–5 mm
  → Use sieving or filtering techniques followed by optical microscopy for preliminary analysis.
- 1 μm–1 mm
  → Use microscopy coupled with spectroscopic methods (e.g. FTIR or Raman or LDIR) for identification and quantification.
- <1 μm (nanoplastics)
  → Proceed with advanced techniques (e.g. SEM/TEM, DLS, or AF4 coupled with spectroscopy).

Fig. 4.1 Decision tree for MP research

Question 2: What instrumentation is available?

Basic Instruments (Microscopes)

- If an optical microscope is available:
  Conduct visual analysis for size, shape, and colour (though not essential).
  → Research Question: Is qualitative data on MPs sufficient?

Spectroscopic Instruments

- ATR FTIR (Attenuated Total Reflectance Fourier-Transformed Infrared Spectroscopy):
  → Analyse the chemical composition of particles >300 μm.
  → Suitable for polymer-type identification and quantification.

- μ-FTIR (Micro-Fourier-Transformed Infrared Spectroscopy):
  → Analyse the chemical composition of particles >20 μm.
  → Suitable for polymer-type identification and quantification.
- μ-Raman Spectroscopy:
  → Use for particles as small as 1 μm.
  → Best for detailed polymer and additive identification.

Electron Microscopy

- SEM/TEM (Scanning or Transmission Electron Microscopy):
  → Analyse surface morphology or ultrastructure for micro and nano-plastics.
  → Coupled with EDX for elemental composition, though limited capabilities for polymer identification.

Chromatographic and Advanced Techniques

- Py-GC-MS (Pyrolysis-Gas Chromatography–Mass Spectrometry):
  → Identify polymer types through thermal degradation. However, this has limited application as there will be no information on the size and number of particles.
  → Suitable for bulk composition analysis.
- AF4-MALS/DLS (Asymmetric Flow Field-Flow Fractionation with Multi-Angle Light Scattering or Dynamic Light Scattering):
  → Analyse nanoplastic size distribution and stability.

Question 3: What is the research question?

MP inventories (larger MPs 1–5 mm)

- Counting techniques with filters and microscopes and hot needle tests.
- Quantify MPs in terms of particles per unit volume or mass.

MP inventories and the polymer types?

- Use FTIR, Raman spectroscopy, LDIR, or Pyrolysis-GC–MS depending on the size range of interest.

What is the potential toxicity of MPs?

- In this case the smaller MPs and nanoplastics will be more important due to their higher surface to volume ratios. Smaller MPs using micro-Raman and NPs using AF4-MALS/DLS or biotoxicity tests with leachate analyses.

How do MPs interact with contaminants?

- Use spectroscopic analysis (FTIR/Raman) combined with adsorption experiments to examine interactions with chemicals such as heavy metals, persistent organic pollutants (POPs), pesticide residues, pharmaceuticals and others.

4.2 Theoretical Basis

The procedure is based on the analysis of MPs in three size fractions: 1–300 μm, 300 μm to 1 mm and 1 mm to 5 mm. The sampling focuses on MPs collection from soil (10–20 cm depth) using a composite sample. The survey design approach allows an unbiased assessment of the target agricultural and non-agricultural soil, irrespective of its size and composition.

The analysis of MPs includes sieving, density separation, the removal of organic matter, and reporting the total count of MPs using a microscope and ATR-FTIR, μ-FTIR, micro-Raman or LDIR.

The method targets synthetic polymers with variable specific densities of <1.8 g cm^{-3}. This is reflected in the use of a saturated solution of NaI (1.8 g cm^{-3}) to separate MPs from the sampled soil. NaI is selected due to its high density. Alternative salts of higher densities may be used for subsequent extractions if users desire and should be reported in the metadata sheet.

After separation by density, the chemical digestion of the supernatant is carried out with H_2O_2 (15–35%) or KOH to eliminate organic matter and reduce its interference in the count of MPs.

4.3 Materials, Reagents, and Equipment

All materials used for sampling, storage, and analysis must be glass or metal.

4.3.1 Equipment and Materials for Fieldwork

- Pencil and logbook (data from the field)
- GPS for georeferencing sampling points (WGS84, coordinates in decimal degrees)
- Batteries (+ extra)
- Camera
- Measure tape at least 20 m long
- A 50 × 50 cm square frame divided into four sections of 25 × 25 cm. The frame can be made either of wood or metal, and metal or wooden stakes can also be used to delimit the sampling area.
- 10 L metal bucket
- Shovel suitable for surface soil collection
- Auger
- Metal or glass containers with lids of at least 2 Kg capacity
- Tape of paper adhesive for sample labelling
- Permanent ink markers
- Aluminium foil

4.3.2 Laboratory Equipment and Materials

- Oven (temperature 200 °C)
- Muffle furnace for cleaning laboratory glassware
- Oven for drying samples (temperature ≤50 °C)
- Analytical balance (sensitivity 0.0001 g)
- Sieve shaker for cascade sieving (suitable for sieves at least 20 cm in diameter and 5 cm in height)
- Magnetic stirrer for solutions
- Vacuum pump
- Stereo or compound microscope or dissection microscope, preferably a UV microscope (minimum 40× and preferably 100x magnification)
- Laminar flow hood (recommended)
- Stainless steel or aluminium trays for drying samples
- Stainless steel sieves (at least 20 cm in diameter and 5 cm high) with 45 μm, 63 μm, 125 μm, 250 μm, 300 μm, 500 μm, 1 mm, 2 mm, 3 mm, and 5 mm metal mesh
- Glass beakers of various volumes (600–5000 mL)
- Aluminium foil
- Watch glass (to cover beakers)
- Magnetic glass stirring bars
- Glass filtration system for 45 mm diameter filters (filtration funnel, base with filter holder and clamps)
- Kitasato flask of at least 1 L
- Metal filters 45 mm diameter and 25 μm pore size
- Fine-tipped antistatic stainless steel twisters
- Pasteur glass pipette
- 76 × 52 × 1 mm slide for microscope
- Paper adhesive tape for sample labelling
- Washing bottle with filtered distilled or deionised water (filtered with 0.45 μm filter)
- Cotton lab coat
- Gloves and safety glasses for preparation and handling of work solutions

4.3.3 Reagents and Solutions

- Water filtered by 0.45 μm or smaller
- NaI (analytical grade)
- Liquid detergent
- Hydrogen peroxide 15–35% (H_2O_2)
- Saturated NaI solution (1.8 g cm^{-3})

4.4 Sample Collection in the Field

Soil, being a non-homogeneous matrix, needs special consideration while sampling. The soil sampling strategy laid down by ISO 18400-102 (2017) is relevant to sampling for MPs in soil. The sampling can be for the hot spots, systematic grid or transect (Piehl et al. 2018), random or stratified (Corradini et al. 2019) very much depending on the research question, whether one is looking at baseline levels in soils or sludge-applied soils or spillover from agricultural farms, wastewater treatment plants, roads, precipitation and even Aeolian deposits.

A critical point to consider is the sampling depths, as they can significantly differ due to management practices such as ploughing (18400-102, I 2017). The minimum sampling depth in Germany is recommended as 30 cm (BBodSchV 1999), whereas in most studies on MPs in agricultural soils, the top 5 cm is sampled (Piehl et al. 2018; Liu et al. 2018).

The USEPA and USDA recommend taking control samples of the same soil type from a pristine area non-affected by the contaminant of concern (Schoeneberger et al. 2012; USEPA 2020). However, because of the omnipresence of MPs and their transport via aerosols with precipitation, it is recommended that random samples of non-agricultural areas away from urban populations should be taken to reduce the risk of bias due to fertiliser, sludge application or carryover.

4.4.1 *General Considerations for Sampling*

Sampling should only be carried out under suitable weather conditions. Sampling should be avoided under heavy rain, immediately after heavy rain, when the fields are flooded and under conditions of heavy dust, gusts, snow cover and frost.

For the purpose of inventories and baseline sampling should ideally cover different soil types and vegetative covers, including bare or sparsely vegetated arable land, low-growth grassland (e.g. shortly after mowing), forest, orchards, etc.

4.4.2 *Site Selection and Description*

- Prior permissions should be obtained (if sampling is done from private property or areas with controlled access).
- Record the land cover, agricultural management practices (ploughing, furrowing), fertiliser application, plastic mulch, pest control, sludge addition, proximity of road and any other relevant information.
- Provide a map excerpt or sketch of the sampling area, including the location of the sampling grid.
- If surface plastic contamination (littering, films, etc.) is visible, document it with photos.

- Check if there are existing data from previous surveys for the site(s) and verify whether they still reflect current conditions during sampling. Any changes observed during site assessment should be noted in the relevant parameters.

4.4.3 On-Site Sampling

- A soil sample with a depth of 10–20 cm should be collected using a stainless steel auger of 3–5 cm diameter, and several samples should be collected to make a composite sample that is more representative of a site and overcome the issue of MP heterogeneity in soils.
- The sampling fields should be divided into three subplots, and a composite sub-sample of 15–30 samples should be drawn from each subplot, covering up to 1 ha. At least three aliquots should be collected from composite samples as replicates.
- Large pebbles, stones and wood pieces should be removed during the sample collection.
- Each sub-sample should be labelled with a legible permanent marker or with a waterproof label on at least two sides of the container.
- However, if the vertical migration of microplastics in soils is to be studied, a core should be collected using a 10 cm steel barrel and sliced at each centimetre.
- In this case, each core should be considered individually.
- The soil samples should be stored in pre-cleaned wide-mouth glass jars or aluminium containers, and initial (wet) weight should be recorded.
- The samples are to be stored at room temperature until further analyses.
- It is recommended that with each sampling campaign, a field background sample be taken from a non-agriculture area away from urban populations and major roads to reduce the risk of bias due to fertiliser, sludge application or carryover.

4.5 Sample Processing

The soil samples are processed using the following approach:

- Fractionation (into three size groups)
- Density separation
- Organic matter removal (digestion with H_2O_2 or KOH)

Note: During the analysis of the samples, special care must be taken regarding the possible contamination from fibres and smaller MPs in aerosols in the laboratory environment. The samples can be classified into three classes: a different instrumentation level and laboratory setup (see decision tree).

1. 1–20 μm
2. 20–300 μm
3. 300–1000 μm

4.5.1 Pretreatment of Soil Samples

- Dry samples in the oven at ≤50 °C to constant weight. (Ambient temperature should be reported in the metadata.)
  Note: To avoid contamination of the samples by MPs in the laboratory environment, cover the sample with aluminium foil during drying. Make some perforations to the aluminium to allow water to evaporate.
- Once samples have dried and reached a constant weight, record the dry weight of each sample in the corresponding data sheet.
- Separately, place each sample in the automatic sieve shaker and cascade sifting using at least 300 μm, 1 mm, and 5 mm sieves for a minimum of 15 minutes. However, we highly recommend using a complete sieve set to provide better resolution for smaller MPs.
- Recover the fractions retained in the base for size fraction (1 μm to <300 μm), sieves of 300 μm (300 μm to <1 mm) and 1 mm (1 mm to <5 mm), and transfer them to respective labelled beakers.
- Discard any material retained on the 5 mm sieves, but take a photo and record the information as metadata.

4.5.2 Density Separation and Extraction of Microplastic Particles

The following steps should be performed separately for each of the size fractions obtained from cascade sieving:

- Add saturated and filtered (0.45 μm) NaI solution (~1.8 g cm^{-3}) to a large glass beaker (1000 ml). Ensure that the solution is about five times the volume of the sample. Add a glass magnetic stirrer.
- Pour the soil fraction into the glass beaker. Pouring in this way allows the particles to start separating. Top up with more NaI solutions if they are insufficient.
- Add a drop of pre-filtered liquid detergent as a surfactant to the mixture to prevent particles from adhering to the walls of the beaker. We recommend using

aqueous sodium hexametaphosphate solution for the samples in a glass beaker (Zhou et al. 2020; Garcés-Ordóñez et al. 2019) to help dissolve aggregates and lumps.

- Cover the glass jar with aluminium foil or watch glass and stir gently on a magnetic stirring plate for 20–30 minutes.
- After agitation, allow the sample to stand overnight.
- Filter the supernatant through a 0.8 µm metal or an anodisc filter (47 mm diameter) using a vacuum filtration system.
- Once the entire sample has been filtered, rinse the walls of the flask or filtration funnel and the filter using pre-filtered distilled/deionised water to collect any remaining particles and eliminate possible traces of salt precipitates that may have formed during the process.
- Stop the pump and let air into the filtration system to remove the vacuum.

Note: If desired by individual laboratories, a second separation with a denser solution could be performed on the remaining sediment. This will allow them to collect supplementary data to align with any other in-house protocols.

4.5.3 *Organic Matter Removal*

For soil samples, the best results for organic matter removal were with H_2O_2 (30–35%). This method is currently recommended for the digestion of organic matter in MP samples (Hurley et al. 2018).

- This reaction should be performed in a fume hood and on ice as this is exothermic.
- Transfer the sample on the filter to 200 ml beaker with a magnetic stirrer.
- The sample should be kept in an ice bath, and the H_2O_2 should be added in a ratio of 1:1 until no bubbles are observed. However, we recommend adding H_2O_2 bit by bit. If the sample has not turned clear, more H_2O_2 may be added.

4.6 Identification and Quantification of Microplastics

Once the sample is density-separated and organic matter is removed, it should be filtered on a silver or anodisc membrane filter, and a multi-tier identification process must be undertaken. If the LDIR is to be used for polymeric characterisation, a gold filter will be used.

4.6.1 Microscopic Identification

- The first step is identification under a microscope, considering the particles that show the absence of any cellular structure, consistency in colour, higher relief, and sharp edges.
  Note: The abundance of natural fibres due to roots and the addition of sewage sludge make it very challenging to differentiate between plastic polymers and matrix interference.
- It is recommended that the particles be marked as tentative MPs after microscopic examination.

4.6.2 Spectroscopic Characterisation

- Looking at the complexity of the chemical composition of MPs and morphology, merely the microscopic examination can not confirm its presence with a high degree of certainty. It is recommended to check 10–20% of the particles on the filter (if they are over 500) and all the particles if they are less than 500 to confirm their synthetic fingerprint (synthetic polymer, plastic) using one of the techniques below.
- The four most commonly employed techniques are:
  The attenuated total reflectance Fourier transformed infrared spectroscopy (ATR-FTIR) is effectively used in analysing MPs of >300 μm here, each single particle is investigated individually.
  The other technique is the micro-Fourier transformed infrared spectroscopy (μ-FTIR), which allows analyses of the MP over the filter sample directly and has an advantage when identifying polar groups. It is often more powerful and can identify MPs up to a finer resolution of 10–20 to 300 μm (Kappler et al. 2016).
  The μ-Raman spectroscopy provides the highest resolution of 1 μm (Uddin et al. 2021) and a better response for non-polar bonds.
  Laser direct infrared (LDIR) chemical imaging system is another very useful technique, but here, the filtration system is a 25 mm apparatus with gold filters; it is very effective for detecting particles >50 μm in size. However, 70–80% of particles are in the unassigned category and require a careful spectral evaluation.
  Note: Another technique that can be quite useful for identifying polymers is the pyrolysis gas chromatography-mass spectrometry (Py-GC–MS). Unlike FTIR and Raman, this destructive technique is quite efficient in identifying polymer types but not in the quantification of MPs in terms of size, shape and number, making any comparison difficult with published and widely reported MP data in soil.

4.7 Reporting

- The abundance of MPs in soil should be reported based on the area or mass, preferably both.
- Determine the abundance of MPs per m².

$$\text{Abundance of MP}\left(\text{MP}/\text{m}^2\right) = \frac{\#\,\text{total of microplastics in the filter}^{(1)}}{\text{Area of sample analyzed}\left(0.0625\text{m}^2\right)}$$

- Determine the abundance of MPs per kilogram of dry soil

$$\text{Abundance of MP}\left(\text{MP}/\text{kg}\right) = \frac{\#\,\text{total of microplastics in the filter}^{(1)}}{\text{Dry weight of the sample }\left(\text{kg}\right)}$$

(1)The number of microplastic particles identified in the procedure target shall be subtracted from the values obtained in the samples of the corresponding batch.

4.8 Quality Control

To minimise sample contamination by MP particles, it is recommended to:

- Carefully clean all working material with running water and detergent, and rinse with filtered distilled water.
- Decontaminate all glass material in a furnace at 500 °C for 4–6 h to eliminate contamination.
- Keep the materials protected with aluminium foil after cleaning until their use to prevent cross-contamination.
- Be careful not to generate cross-contamination during sampling and preparation due to using the same material (shovels, buckets, etc.).
- Thoroughly clean the work area with 70% technical alcohol so it is free of dust or microplastic particles.
- Avoid having multiple personnel in the work area during the testing process.
- Avoid using synthetic clothing; use 100% cotton lab coats or easily distinguishable colours, such as orange and green, used in traffic uniforms.
- Clean metal filters with 100% ethanol, and verify under stereomicroscope before use.
- Avoid exposing samples to air more than necessary.
- Ideally, sample processing should be performed in laminar flow hoods.
- All working solutions must be filtered with a 0.45 μm pore size filter before use.

4.8.1 *Blank Analysis*

- Running a procedural blank using a 100-gram soil sample baked at 500 °C for 4–6 hours and 1 L of filtered water is recommended.
- The blanks should be subjected to the entire analytical process together with the samples.
- At least one procedure blank shall be analysed for each batch of 10 samples maximum, i.e. if the working batch is greater than 10 samples, a second blank must be performed.

4.8.2 *Positive Controls or Standard Reference Material*

- It is recommended to spike a soil sample that was baked at 500 °C with the known number and type of polymers.
- With each batch at least one Positive control should be analysed
- If standard reference material (SRM) is accessible, it can be used instead of a positive control for each batch of analyses.

4.8.3 *Repeat the Microplastic Detection and Counting Process*

- The counting of MPs on the filter should be repeated by a second analyst.
- Compare the results obtained on a randomly chosen filter.
- Laboratories should determine the coefficient of variation of possible MPs identified by microscopic examination by different operators, and the variance should be reported. We recommend that 10% be considered the maximum for accepting the results.
- If there is over 10% variation, it should be given to the third operator and fourth.

4.8.4 *Proficiency Test*

- Laboratories are highly encouraged to participate in proficiency tests to evaluate the analytical performance of the laboratory.

Acknowledgements Authors are thankful to the Kuwait Institute for Scientific Research for supporting the study under project EM127C and to the International Atomic Energy Agency for the award of the research contract No. 26641 entitled "Assessment of Microplastic Input and Transfer to Agricultural Soils by Sewage Sludge Application" under the Coordinated Research Project (CRP): "Assessing the Fate, and Environmental Impact of Plastics in Soil and Crop Ecosystems Using Isotopic Techniques (D15021)."

References

18400-102, I (2017) Soil quality-sampling-part 102: selection and application of sampling techniques. International Organization for Standardization, Geneva, Switzerland. Technical Report

BBodSchV (1999) Federal soil protection and contaminated sites ordinance; number on the basis of 6, 8 paragraphs 1 and 2 and § 13 paragraph 1 sentence 2 Federal Soil Protection Law of 17 March 1998. Federal Law Gazette Bundesgesetzblatt: Berlin, Germany. I p 502

Corradini F et al (2019) Evidence of microplastic accumulation in agricultural soils from sewage sludge disposal. Sci Total Environ 671:411–420

Gao N et al (2020) Thermochemical conversion of sewage sludge: a critical review. Prog Energy Combust Sci 79:100843

Garcés-Ordóñez O et al (2019) Marine litter and microplastic pollution on mangrove soils of the Ciénaga Grande de Santa Marta, Colombian Caribbean. Mar Pollut Bull 145:455–462

Hurley RR et al (2018) Validation of a method for extracting microplastics from complex, organic-rich, environmental matrices. Environ Sci Technol 52(13):7409–7417

Kappler A et al (2016) Analysis of environmental microplastics by vibrational microspectroscopy: FTIR, Raman or both? Anal Bioanal Chem 408:8377

Liu M et al (2018) Microplastic and mesoplastic pollution in farmland soils in suburbs of Shanghai, China. Environ Pollut 242:855–862

Piehl S et al (2018) Identification and quantification of macro- and microplastics on an agricultural farmland. Sci Rep 8(1):17950

Schoeneberger P et al (2012) Natural resources conservation service. National Soil Survey Center, Lincoln. Field Book for Describing and Sampling Soils, Version 3.0

Uddin S et al (2021) A review of microplastic distribution in sediment profiles. Mar Pollut Bull 163:111973

USEPA (2020) LSASD Operating Procedure for Soil Sampling; Technical Report LSASDPROC-300-R4. Laboratory Services and Applied Science Division, AthensGA

Weber CJ, Santowski A, Chifflard P (2022) Investigating the dispersal of macro- and microplastics on agricultural fields 30 years after sewage sludge application. Sci Rep 12(1):6401

Zaman M, Uddin S (2024) Sewage sludge usage in agriculture: are we introducing microplastics in food chain? Soils Newslett, International Atomic Energy Agency 10(2):7–8

Zere Taskin S, Bilgili U (2023) Using sewage sludge as alternative fertiliser: effects on turf performance of Perennial Ryegrass. Sustainability 15(18):13597

Zhou B et al (2020) Microplastics in agricultural soils on the coastal plain of Hangzhou Bay, east China: multiple sources other than plastic mulching film. J Hazard Mater 388:121814

Zubris KAV, Richards BK (2005) Synthetic fibers as an indicator of land application of sludge. Environ Pollut 138(2):201–211

Chapter 5
A Critical Perspective on Nanoplastics in Soil and Crop Systems

Saif Uddin, Mohammad Zaman, Talat Saeed, Antony Andrady, and Montaha Behbehani

Abstract Nanoplastics have emerged as a significant emerging contaminant of the Anthropocene. The ubiquitous presence of nanoplastics attracted considerable attention due to their potential impacts on soil and crop systems. The nanoplastics can be primary and secondary in origin. Their interaction and fate in soil depend on the physicochemical properties, including size, shape, surface charge, hydrophobicity and polymer type. Nanoplastics can both adsorb and leach contaminants, altering soil chemistry and increasing the bioavailability of toxic substances to plants and microorganisms in the rhizosphere. Nanoplastics interferes with soil-plant-microbe interactions, often bringing changes in rhizosphere microbial community composition and function, negatively impacting soil health and crop productivity. Experimental evidence on nanoplastics uptake by roots exacerbates their accumulation in other tissues, including edible plant parts, posing risks to food safety and human health.

Nanoplastics are reported to induce oxidative stress in plants and soil microorganisms, impairing key physiological processes and reducing crop yields. Reports of nanoplastic bioaccumulation within the food chain underscore the need for risk assessments. Despite a growing body of evidence, significant knowledge gaps remain, especially in the detection, quantification, and characterisation of nanoplastics in the environment. Even though pyrolysis GCMS can help identify the polymer, the approach provides no information on the shape, number and size of the nanoplastics, which is a severe constraint in assessing nanoplastic contamination

S. Uddin (✉) · T. Saeed · M. Behbehani
Environment and Life Sciences Research Centre, Kuwait Institute for Scientific Research, Safat, Kuwait
e-mail: sdin@kisr.edu.kw

M. Zaman
Joint FAO-IAEA Centre of Nuclear Techniques in Food and Agriculture, Department of Nuclear Sciences and Applications, International Atomic Energy Agency, Vienna, Austria

A. Andrady
Department of Chemical and Biomolecular Engineering, North Carolina State University, Raleigh, NC, USA

S. Uddin, M. Zaman (eds.), *Micro- and Nano-plastics in Soil and Crop Systems*, Sustainability Sciences in Asia and Africa, https://doi.org/10.1007/978-981-95-2740-3_5

and toxicity, which is size dependent. Significant efforts are needed before any realistic ecological impact assessment can be undertaken to understand their long-term environmental and health implications better.

Keywords Pyrolysis-GCMS · Nanoplastics · Oxidative stress · Rhizosphere

5.1 Introduction

With the progress in research on microplastics (<5 mm in size), there are increasing efforts to understand the fate of even smaller plastic particles: the nanoplastics (Gigault et al. 2016). The upper size limit of nanoplastics is still being discussed, with the literature variously suggesting either 100 nm or 1000 nm (Gigault et al. 2018a, b; Li et al. 2020a; Alexander et al. 2016; Ng et al. 2018). A reliable report of the likely presence of nanoplastics in the environment was first put forward by Gigault et al. (2016) suggesting that microplastics can fragment under solar radiation. The presence of nanoplastics within soil and crop systems has emerged as a critical issue (Atugoda et al. 2023) as they potentially present a more insidious form of contamination for multiple reasons, i.e. their enhanced toxicity, their colloidal behaviour and stability under a variety of environmental conditions, ability to move across cell membranes, and most importantly, the difficulties in isolating and identifying them in environmental samples.

Due to small size and unique physicochemical properties (Gigault et al. 2021; Covernton et al. 2019), nanoplastics pose distinct challenges to soil health (Wu et al. 2020; Hurley and Nizzetto 2018), crop productivity, and human health. They can originate from a variety of sources, both primary and secondary. Primary nanoplastics are intentionally manufactured at the nanoscale for use in industrial applications (Stephens et al. 2013; Rodríguez-Hernández et al. 2020), including abrasives, inclusions in cosmetics (Hernandez et al. 2017), fillers in coatings, and drug delivery systems (Marchesan and Prato 2013; Shi et al. 2015; Ali et al. 2020). These particles often enter soil systems through wastewater irrigation, land application of sewage sludge (Frehland et al. 2020), flooding, and atmospheric deposition. Secondary nanoplastics, on the other hand, are formed through the degradation of larger plastic debris in the environment (Gigault et al. 2021), and one of the conspicuous sources of smaller plastics is mulch (Seo et al. 2024; Nizzetto et al. 2016; Bao et al. 2024; Ferreira-Filipe et al. 2024). Photodegradation, mechanical weathering, and biological activity (Cai et al. 2023) can break down macroplastics and microplastics into nanoplastic particles (Wang et al. 2020; Astner et al. 2019; Zhang et al. 2019). Agricultural practices, such as using plastic mulching films, fertilisers, and pesticide containers, contribute significantly to soil nanoplastics accumulation. Based on the 88–99.9% of microplastic retained in sewage sludge (Uddin et al. 2020, 2022), it can also be a significant source of nanoplastics. However, data on nanoplastic inventories are scarce due to challenges such as matrix interference in their analysis; sewage sludge is rich in organic compounds, dissolved solids and microorganisms (Cavazzoli et al. 2023).

Two agricultural practices are primarily responsible for introducing micro- and nanoplastics into agricultural soil: The pervasive use of plastics in agriculture exacerbates this issue. For instance, plastic mulches, widely used to conserve moisture and control weeds, degrade over time, releasing microplastics and eventually nanoplastics into the soil (Atugoda et al. 2023; Astner et al. 2019; Zhang et al. 2019); and the application of biosolids as fertilisers that transfer the NPs present in sewage sludge directly into agricultural soils (Mclaughlin and Pennock 2018).

5.2 Behaviour of Nanoplastics in Soil Systems

The behaviour of nanoplastics in soil is influenced by their physicochemical properties, including size, shape, surface charge, and hydrophobicity. Due to their small size, nanoplastics have a large surface area-to-volume ratio, which enhances their reactivity and interaction with other soil components. Nanoplastics can adsorb metals on the surface and absorb persistent organics in the bulk (Rochman et al. 2013a, b, 2014), potentially altering the chemical properties of the soil. Additionally, their hydrophobic nature can influence soil water retention and transport processes, affecting soil structure and function.

The interaction of nanoplastics with the microorganisms in soil can result in changes within the microbial community composition and function (Aralappanavar et al. 2024). These interactions can have far-reaching effects on soil fertility and health, as the soil microbes play a pivotal role in maintaining soil structure, nutrient cycling, and decomposition of organic matter (Aralappanavar et al. 2024; Nath et al. 2024). The vertical migration of nanoplastics with the irrigation return flow or infiltration with precipitation can transport them to deeper soil layers and even to underground aquifers. The mobility of nanoplastics in soil is influenced by several factors, such as soil texture, moisture content, and organic matter. Understanding the fate and transport of nanoplastics in soil is essential for assessing their long-term environmental impacts.

Extracting nanoplastics, especially from organic-rich soils, is challenging (Wahl et al. 2021), hence the scarcity of nanoplastic inventories in soils (Hurley and Nizzetto 2018). A detailed review by Wahl et al. (2021) on agricultural soil in France that has been contaminated with plastic debris for about 30 years appears to be a proof of concept, but lacks an inventory of the site. They extracted nanoparticles from soil samples using soil-water extraction with ultra-pure water in a 1:4 ratio, stirred at 300 rpm for 72 hours without any pH adjustment (Wahl et al. 2021). The samples were filtered on 0.8 µm cellulose acetate filters (Wahl et al. 2024), and the water extract was fractionated utilising asymmetric flow-field flow fractionation (AFF) where the accumulation wall was a polyethersulfone (PES) membrane. The use of AFF coupled online to a static light scattering (AF4-UV-SLS) in the study suggests the colloidal populations in two size ranges of 20–150 nm and 150–500 nm, whereas the Py-GCMS was used to identify polymers.

As in the French agricultural soil amended with sewage sludge, NPs are likely ubiquitous in other regions' agricultural farms and soils. In another study, nanoplastics were reported in the soil profile 0–55 cm (Wahl et al. 2024). Ng et al. (2018) computed a lifetime addition of 2.3–63 tons per hectare of microplastics to agricultural soils, which Atugoda et al. (2023) misattributed as a nanoplastic and an annual addition rather than a lifetime. Nevertheless, microplastic degradation may result in nanoplastic. Still, its quantification is impossible as the nanoplastic formation rate will depend on the type of polymer, soil type and conditions favourable to microbial degradation, such as organic matter and moisture content. The microplastic inventories were higher in the topsoil, i.e., 78.00 items kg^{-1} compared to 62.50 items kg^{-1} in deep soils (Liu et al. 2018a). However, Liu et al. (2018a) also indicated that the larger plastics were found in the topsoil while the deeper soils were enriched with smaller microplastic size fractions; with this consideration, should we expect nanoplastics inventories to be higher in the soil beneath? This can partly be due to the movement of NPs through the interstitial spaces in the soil profile, with the irrigation return flow.

5.3 Impact on Crop Systems

The presence of nanoplastics in soil can directly and indirectly affect crop systems. One primary concern is plants' potential uptake of nanoplastics (Li et al. 2021a). The first evidence of nano bead internalisation in crops came in 2012 (Bandmann et al. 2012), where it was reported that 20 nm beads were taken up quickly by BY-2 cells, while the 100 nm size beads were excluded from uptake by turgescent and plasmolysed cells (Bandmann et al. 2012). On the other hand, the protoplasts could internalise 1000 nm size beads. Studies have shown that nanoplastics and smaller MPs can be absorbed by plant roots (Dong et al. 2021; Li et al. 2021b; Liu et al. 2022) and translocated to various plant tissues, including leaves (Dong et al. 2021; Sun et al. 2021), stems (Liu et al. 2022; Sun et al. 2021; Lian et al. 2020a), and fruits (Li et al. 2020b, 2021b).

This uptake can occur through various complex mechanisms and pathways, including uptake by root pores from soil colloids and endocytosis. According to a study (Bandmann et al. 2012), bead internalisation was only partially hindered by Ikarugamycin, a clathrin-specific inhibitor, while wortmannin completely blocked it. This suggests that endocytosis happens through both clathrin-dependent and independent pathways, supporting the view that clathrin-independent endocytosis is widespread in plants.

The distribution and movement of nanoplastics (NPs) in soil are significantly influenced by soil colloidal behaviour, which governs their transport, retention, and aggregation (Liu et al. 2021a). Soil colloids, including clay minerals, organic matter, and metal oxides, interact with NPs through electrostatic forces, van der Waals interactions, and cation bridging, affecting mobility (Adeel et al. 2021). Nanoplastics adsorb onto colloids, forming heteroaggregates that can lead to immobilisation as larger aggregates are filtered out in the soil matrix (Gigault et al. 2018b; Pradel et al. 2023). Furthermore, environmental factors such as soil pH, moisture content, ionic strength and organic matter levels influence colloidal stability; stable colloids can transport attached nanoplastics deeper into the soil profile, potentially reaching groundwater, while conditions promoting flocculation can result in surface deposition (Xu et al. 2023a). Thus, the dynamic interplay between soil colloidal behaviour and nanoplastic properties is critical for understanding contaminant dispersion (Pérez-Reverón et al. 2023), with significant implications for environmental risk assessment and management practices.

The translocation of nanoplastics within plants raises concerns about their accumulation in edible plant parts, potentially entering the food chain and posing risks to human health. In a study, peat moss (*Sphagnum palustre*) exposed to 50 nm polystyrene beads demonstrated preferential adsorption of large aggregates on the leaf surface. Additionally, single nanoparticles were localised within the hyalocysts, which are empty cells containing pores ranging from 5 to 10 μm in diameter and in the cytoplasm (Capozzi et al. 2018). Nanoplastics ranging in size from 50 to 500 nm were also detected in garden cress (*Lolium perenne*). These nanoparticles obstructed the seed capsule's pores, with a pronounced accumulation on the root hairs (Bosker et al. 2019). These particles were also detected on the leaves and epidermal surfaces (Bosker et al. 2019). In another study, 20 nm polystyrene beads were detected within the midrib and veinlets of *Vigna radiata* (mung bean), with prominent aggregation observed in the veins—indicating nanoplastic translocation from the soil into the plant body (Chae and An 2020). Additionally, nanoparticles were observed along the vessel walls in the exposed leaves (Chae and An 2020); however, the possibility of atmospheric deposition and endocytosis cannot be ruled out.

In another experimental study, 50 nm and 500 nm polystyrene beads were applied to soil planted with *Spirodela polyrhiza* (great duckweed). The nanoplastic beads were predominantly aligned along the root shafts and tips, with a uniform distribution observed across the entire root surface (Dovidat et al. 2020), suggesting their likely uptake into plants. In another experiment using *Allium cepa* (garden onion), uptake of 50 nm polystyrene beads was reported (Giorgetti et al. 2020), where the nanoplastics accumulated within the cellular compartments, including vacuoles and the cytoplasm of the root epidermis, resulting in a reduction in root growth.

The genotoxicity of 100 nm polystyrene beads in *Vicia faba* (Faba bean) was reported; similar to great duckweed, the nanoplastics were found localised within the root tips (Jiang et al. 2019) due to either endocytosis or colloidal transfer, elucidating the potential translocation of nanoplastics from the soil into the plants (Jiang et al. 2019). *Triticum aestivum* (wheat) was exposed to 200 nm and 2.0 μm polystyrene beads in another experimental study. These beads were detected in the stems and leaves, providing evidence of the ability of nanoplastics to move through the plants (Lian et al. 2020b). In an additional experiment by the same group, both *Triticum aestivum* and *Lactuca sativa* (lettuce) were treated with 200 nm and 2.0 μm beads composed of polystyrene and poly(methyl methacrylate) (PMMA); nanoplastics were detected in the roots, shoots, xylem, and leaf veins, further indicating effective translocation of nanoplastics from soil to the crop. The presence of nanoplastics within different parts of the crops underscores their ability to impact both plant physiology and food safety.

In another study, the barley roots readily absorbed PMMA beads of 96.75 ± 0.58 nm diameter (Li et al. 2021c), suggesting the capability of nanoplastic incorporation into plant roots and likely influence in nutrient uptake and effect on plant health. Studies have shown different behaviours of positive and negative charges on 200 nm polystyrene beads in *Arabidopsis thaliana* (Arabidopsis). The negatively charged beads were mainly localised within the stele of the maturation zone, root tips, and the border cells of the root (Sun et al. 2020). Meanwhile, the positively charged beads were on the root hairs and within intercellular spaces. These results underpin the importance of surface charge in controlling the nanoplastic uptake and distribution in plant tissues. In both *Triticum aestivum* (wheat) and *Arabidopsis thaliana* (Arabidopsis), exposure to 40 nm and 1000 nm polystyrene beads resulted in a pronounced accumulation along the root surface, with a particularly notable concentration at the root tip (Taylor et al. 2020). This evidence, along with other reports, suggests that root tips are the main point of uptake of nanoplastics. The studies included here explicitly highlight the importance of particle size in adhesion uptake and endocytosis, providing significant insights towards the mechanisms of nanoplastic mobilisation from the soil into monocot and dicot plant species. Evidence of enhanced antioxidant activity and activated carbon metabolism has been reported due to the localisation of 20 nm polystyrene beads in the roots of *Oryza sativa* (rice) (Zhou et al. 2021).

Based on the published information, it can be hypothesised with reasonable confidence that nanoplastics in the rhizosphere (Chen et al. 2022a) can impact plant growth and development. The nanoplastics in the soil region surrounding plant roots can alter root morphology, affecting plant growth (Roy et al. 2022). The nanoplastic exposure impedes root elongation and affects root hair growth and nutrient uptake (Khan et al. 2024; Hussain et al. 2025). Such physiological changes reduce overall plant health and lower crop yields (Jia et al. 2023). Nanoplastics are also known to affect soil-plant-microbe interactions, potentially disrupting beneficial symbiotic relationships, such as those with mycorrhizal fungi and nitrogen-fixing bacteria.

Moreover, nanoplastics can interact with other soil contaminants like metals and organics, potentially enhancing their bioavailability and toxicity to plants. For example, nanoplastics can adsorb heavy metals (Xu et al. 2023b; Ceylan et al. 2024) and organic pollutants (Liu et al. 2021a), sorbed by nanoplastics, may facilitate their uptake by plant roots and increase their accumulation in plant tissues (Chen et al. 2022b). This co-contamination can have synergistic effects, exacerbating the toxic effects of pollutants on crops and compromising food safety (Liu et al. 2021b).

5.4 Ecotoxicological Concerns

With the extremely large specific surface area, nanoplastics release additives that are added to plastics manufacturing at a faster rate. The lipophilic additives on leaching can penetrate cell membranes and potentially bioaccumulate (Kwon et al. 2017). Additionally, residual monomers from polymerisation in producing plastics, solvents, and additives may leach out of the plastic matrix, posing chronic toxicological risks to plants and organisms (Atugoda et al. 2023; Omidoyin and Jho 2024; Ramanayaka et al. 2024). There is an immediate need to assess the leaching rates of these additives from different polymers under ecologically relevant environmental conditions; the available information is not comprehensive enough to evaluate the actual risk it may pose.

Hydrophobic contaminants like hydrocarbons, polycyclic aromatic hydrocarbons (PAHs), polybrominated diphenyl ethers (PBDEs), polychlorinated biphenyls (PCBs), and pharmaceuticals (Shen et al. 2019) in the soil or water are adsorbed over nanoplastics due to their high partition coefficient (Sun et al. 2023).

There are only seven studies published on the adsorption of organic contaminants onto nanoplastics of different sizes (Table 5.1). Studies have indicated that PAHs and PBDEs have 2–3 orders of magnitude higher partitioning onto polystyrene nanoplastics than PS microplastics (Velzeboer et al. 2014; Liu et al. 2016); this can be logically attributed to the higher specific surface area-to-volume ratio. In another study, the hydrophobicity of contaminants is stated as a key factor in determining the binding ability of non-polar and weakly polar contaminants to nanoplastics (Liu et al. 2018b), controlling their distribution compared to polar compounds. On the other hand, two chemicals, PCB-77 and PAH benzo[k]fluoranthene, have similar hydrophobicity, but the sorption on PS nanoplastics is much higher for benzo[k]fluoranthene and is explained by stronger π-π interactions (Liu et al. 2016). Higher adsorption of ciprofloxacin (0.41 mg g^{-1}) and bisphenol-A (7.45 mg g^{-1}) on aged nanoplastics (Xiong et al. 2020) indicates enhanced affinity of these contaminants to weathered nanoplastics. The sorption of pharmaceuticals onto nanoplastics can enhance these contaminants' spatial disposition and bioavailability. This co-transport facilitates the widespread distribution of pharmaceuticals beyond their original discharge points, increasing environmental persistence (Alimi et al. 2018). In another study, divalent ions are shown to reduce the sorption of tetracycline on polystyrene nanoplastics by blocking binding sites on the nanoplastic surface, thus preventing interaction with antibiotic molecules (Wan et al. 2019). The sample matrix, pH, salinity, and organic matter influence nanoplastic interaction with contaminants.

A few studies have reported the adsorption of metals onto nanoplastics. Davranche et al. (2020) found high Pb(II) levels on nanoplastics processed in the laboratory from plastic particles extracted from the North Atlantic Gyre. Eleven studies have examined nanoplastic and metal/metalloid effects on the end receptors, including terrestrial and non-terrestrial biota. The studies summarised in Table 5.2 demonstrate the potential for nanoplastics to interact synergistically with various

Table 5.1 Organic Contaminants Adsorbed on nanoplastics of different sizes

Polymer	Size in nm	Contaminants	Adsorption mg/g	References
Polystyrene	70	Polychlorinated biphenyls 180	6.61×10^{133}[a, b]	Velzeboer et al. (2014)
		Polychlorinated biphenyls 126	2.69×10^{76}[b]	
		Polychlorinated biphenyls 77	1.05×10^{31}[b]	
		Chrysene	6.8[a, b]	Liu et al. (2016)
		Benzo[a]pyrene	33.8[b]	
		Benzo[b]fluoranthene	42.6[b]	
		Benzo[k]fluoranthene	33.1[b]	
		Benzo[a]anthracene	7.9[b]	
		Benzo[g,h,i]perylene	20.0[b]	
	50	Phenanthrene	2[b]	Ma et al. (2016)
	300–400	Tetracycline	44.8[b]	Wan et al. (2019)
Polystyrene (Pristine)	40	Ciprofloxacin	5.6[b], 3.34[c]	Xiong et al. (2020)
		Bisphenol-A	1.26[c]	
Polystyrene (Aged)		Ciprofloxacin	5.4[b], 3.75[c]	
		Bisphenol-A	8.71[c]	
Polystyrene, humic acid, pH6	300–400	Tetracycline	51.0[b]	Wan et al. (2019)
Polystyrene, $MgCl_2$, pH6			37.5[b]	
Polystyrene-COOH	200	Ciprofloxacin	5.2[b]	Yilimulati et al. (2021)
	500		4.3[b], 3.82[c]	

[a]$(mg\ kg^{-1}/\mu g\ L^{-1})^n$
[b]Isotherm modelling
[c]Kinetic modelling

metals, leading to increased toxicity in several organisms. Many of these organisms are not in the soil, but these findings underscore the importance of considering the combined effects of nanoplastics and metals. The paucity of soil data will limit any realistic environmental risk assessments, as their co-exposure can significantly amplify harmful outcomes among the soil microbial community and probably can also affect the crops.

The ecotoxicological impacts of nanoplastics on soil and crop systems are not yet fully understood, but emerging evidence suggests that they can potentially pose significant risks to soil microbiota and crops. Nanoplastics can induce oxidative stress in plants by producing reactive oxygen species (ROS) that cause cellular damage. This stress can impair photosynthesis, reduce growth, decrease crop productivity, affecting germination, flowering, and fruiting in plant's. The potential for nanoplastics to accumulate in the food chain is also a realistic concern. If nanoplastics are taken up by crops, they could be transferred to herbivores and, subsequently, to higher trophic levels, including humans. The long-term health implications of

Table 5.2 Experimental data on polystyrene nanoplastic and metal toxicity on biota

Size in nm	Metal	Significant findings from treatments of coexposure	References
10	Cadmium	The concurrent exposure to nanoparticles (NPs) and cadmium (Cd) markedly accelerated mortality in *Ctenopharyngodon idellus* (grass carp), with varying degrees of tissue damage observed in the gills, liver, kidneys, and muscles—most severe at concentrations of 20 mg L^{-1} Cd combined with 300 μg L^{-1} NP. Additionally, treatment with NPs alone at 30 μg L^{-1} led to increased superoxide dismutase (SOD) activity in both the kidney and liver. In contrast, peroxidase (POD) activity was elevated exclusively in the kidney.	Chen et al. (2022c)
100, spherical		At 0.05 mg Cd L^{-1} there was a significant reduction in *Euglena gracilis* growth; coexposure enhanced microalgal toxicity with synergistic effects. Metabolic disorders and oxidative stress are the main toxic mechanisms.	Cao et al. (2022)
500, spherical	Copper	Significant increase in toxicity compared to nanoplastic or Cu alone oxidative stress, cytoplasmic vacuolation and destruction of chloroplasts in *Chlorella* sp. TJ6-5 and *Pseudokirchneriella subcapitata* NIES-35	Wan et al. (2021)
87.17–106, spherical		No adsorption of copper ions was detected on PS-COOH nanoparticles. However, changes in their surface charge and hydrodynamic diameter indicate the formation of an eco-corona and a subsequent loss of the cell's native morphology at an exposure level of 10 mg PS L^{-1} in the presence of copper. Furthermore, the nanoparticles' ability to interact with exudates from *Raphidocelis subcapitata* and adhere to its cell wall suggests that their transfer through the food chain is plausible.	Bellingeri et al. (2019)
50 & 55 spherical		Exposure suppressed the growth of *Platymonas helgolandica* var. *tsingtaoensis*, intensifying inhibition as exposure duration and concentration increased. Oxidative damage was evident at PS-NP concentrations exceeding 50 mg L^{-1} and Cu levels of at least 1.5 mg L^{-1}, with the combined exposures exhibiting a synergistic effect.	Gao et al. (2022)
100, spherical	Zinc	At environmentally relevant concentrations, nanoplastics and zinc do not exhibit a combined interactive effect on freshwater microeukaryotic communities; however, zinc alone induces a discernible alteration in community structure. In contrast, nanoplastics appear to have no significant impact. Zinc exposure leads to a marked initial reduction in phototrophic microorganisms, though photosynthetic activity eventually recovers, accompanied by a shift in the dominant taxa from bacillariophytes to chlorophytes.	Sieber et al. (2022)
	Arsenic	Consequent to non-environmental concentrations (1 mg/L), there was an interactive effect of nanoplastics and arsenic; fresh weight and growth rates in *Myriophyllum verticillatum* were significantly reduced. However, the negative effects at lower concentrations were mainly due to arsenic exposure; *Myriophyllum verticillatum* responded to stress by regulating amino acids and organic acids.	Li et al. (2023)
		The synergistic effect of nanoplastics and arsenic was higher than their single toxicity on *Vallisneria denseserrulata* and *Potamogeton crispus*. The nanoplastics were observed in the leaf cells of these submerged macrophytes.	Tang et al. (2022)
	Lead	In zebrafish intestinal cells, co-exposure to nanoparticles at 20 and 200 μg L^{-1} alongside 50 μg L^{-1} lead demonstrated that the 200 μg L^{-1} nanoparticle treatment had the most pronounced impact on macrophages, whereas lead exposure predominantly affected enterocytes.	Yu et al. (2022)
50, 200, 500 Spherical	Gold	Smaller nanoparticles were observed to penetrate both developing zebrafish embryos and their chorion. Exposure to nanoplastics resulted in modest alterations in survival, hatching rates, developmental abnormalities, and cell death. The synergistic effects were notably more pronounced with increasing dose and decreasing particle size. Signs of oxidative stress and mitochondrial damage were also reported.	Lee et al. (2019)
1000 spherical	Chromium	NP ~1820/ml and Cr(VI) at 2 and 5 ppm for 72 h showed low mortality of *Daphnia magna* at NP13.3%, 2 ppm Cr(IV) individually, but the synergistic effect of two led to 74.5 mortality, and when Cr concentration was 5 ppm absolute mortality was observed; Microbial diversity was changed significantly.	Varg et al. (2022)

chronic exposure to nanoplastics through food consumption are still largely unknown, but the potential for bioaccumulation and biomagnification warrants serious consideration (Petersen et al. 2023; Miller et al. 2020; Gupta et al. 2022).

5.5 Knowledge Gaps, Research Needs and Concerns

Despite the growing body of evidence on the presence and effects of nanoplastics in soil and crop systems, significant knowledge gaps remain. The lack of standardised methods to detect and quantify nanoplastics in environmental matrices hinders our ability to accurately assess their prevalence and impacts. The method used for NP extraction from soil samples is usually soil-water extraction. It is reported that even the least contaminated water source (Milli-Q water) has 4.7–64.4 MP particles ml^{-1} (Jones et al. 2024). This significant contamination can be introduced during sample processing, and most published reports have not accounted for such contaminations. The samples are then segregated by filtration, often on the filter with nanosize pores; usually, 0.8 μm pore filters have been used. The efficacy of removing NPs from these filter membranes is not apparent, especially as the filter material is not always inorganic or metallic. The technique primarily used for identifying and characterising NPs is pyrolysis gas chromatography with mass spectrometry (pyro-GCMS). The analyses by pyro-GCMS provide pyrolyzate markers for different polymers (Table 5.3), which have been defined in detail by Matsui et al. (2020), where the m/z signal and retention index (RI) for each polymer product were taken into consideration and summated mass spectra (SMS) were constructed. The rationale for using the retention index for peak identification emanates from the fact that, unlike retention time, which depends on the length of the column, thickness of film, diameter, and size of the transfer line (Melo et al. 2017), the RI is independent of the GC parameters. There are discussions regarding the non-inclusion of styrene since pyrolysis of PS and ABS results in its detection; hence, it should not be used as a characteristic product for identifying polymer species.

The complex interactions between nanoplastics and soil constituents, including organic matter, minerals, and microbes, are not yet fully understood. Long-term studies also need to evaluate the chronic effects of nanoplastics on soil health, crop productivity, and food safety. Research on the potential for nanoplastics to interact with other environmental stressors like thermal and microbial degradation, and changes in soil pH, is vital. Developing sustainable agricultural practices that minimise plastic use and reduce nanoplastics input into the terrestrial environment is crucial for mitigating their impacts.

A significant limitation of the pyro-GCMS is its inability to provide information on the size, count, and shape of the nanoplastics in a sample. The nanoplastic size is critical in assessing their toxicity; the higher surface-to-volume ratio of smaller particles has increased potential toxicity. The smaller nanoplastics have enhanced capability to penetrate biological membranes (Khan et al. 2024), but they also translocate and accumulate in tissues away from their point of contact or entry, leading

Table 5.3 The pyrolysis product of different polymers, with their m/z signals and retention index (After Matsui et al. (2020))

Pyrolyzate	Retention index	m/z	Polymer
1-Undecene	1091	41, 55, 70, 97, 111, 154	Polyethylene
1-Tetradecene	1393	41, 55, 97, 125, 196	
1-Eicosene[a]	1994	43, 55, 83[a], 97, 280	
2,4-Dimethyl-1-heptene[a]	843	39, 41, 43, 55, 70[a], 83, 126	Polypropylene
2,4,6-Trimethyl-1-nonene	1082	41, 43, 55, 69, 83, 111, 125, 168	
2,4,6,8-Tetramethyl-1-undecene (isotactic)	1311	41, 43, 55, 69, 111, 125, 154, 167	
2,4,6,8-Tetramethyl-1-undecene (syndiotactic)	1330	41, 43, 55, 69, 111, 125, 154, 167	
3-Butene-1,3-diyldibenzene	1761	65, 91, 104, 115, 130, 193, 208	Polystyrene
5-Hexene-1,3,5-triyltribenzene[a]	2514	65, 91[a], 117, 194, 207, 312	
Styrene monomer	8.5 min[b]	104	
Toluene	7 min[b]		
Benzene[a]	660	39, 50, 51, 52, 77, 78[a]	Polyvinyl chloride
Naphthalene	11.3 min[b]	128	
naphthalene-1-methyl	12.3 min[b]	142	
Methyl methacrylate[a]	711	39, 41, 59, 69, 99, 100[a]	Poly(methyl methacrylate)
ε-Caprolactam[a]	1276	30, 42, 55, 56, 84, 85, 113[a]	Nylon 6
N-(5-Cyanopentyl)hex-5-enamide	1943	30, 41, 55, 69, 100, 114, 154, 168, 208	
6-Acetamido-N-(5-cyanopentyl) hexanamide	2660	30, 43, 55, 114, 167, 224, 267	
Cyclopentenone[a]	799	39, 41, 42, 55, 56, 84[a]	Nylon 66
Adiponitrile	1200	39, 40, 41, 51, 52, 54, 55, 68	
1,8-Diazacyclotetradecane-2,7-dione	2328	30, 41, 55, 98, 112, 140, 198, 226	
4-Vinylcyclohexene	841	39, 54, 79, 80, 93	Acrylonitrile butadiene styrene
2-Methylene-4-phenylbutanenitrile	1363	39, 65, 91[a], 92, 157	
4-Phenylpent-4-enenitrile	1446	77, 91, 103, 115, 117, 130, 142, 157	
2-Methylene-4-phenylheptanedinitrile	1873	91, 104, 117, 144, 182, 210	
2-Phenethyl-4-phenylpent-4--enenitrile[a]	2234	91[a], 105, 115, 118, 156, 170, 261	

(continued)

Table 5.3 (continued)

Pyrolyzate	Retention index	m/z	Polymer
Dimethyl terephthalate[a]	1524	50, 76, 103, 120, 135, 163[a], 194	Polyethylene terephthalate
Bisphenol A dimethyl ether[a]	2110	133, 211, 241[a], 242, 256	Polycarbonate
4-(4-Aminobenzyl)-N-methylaniline[a]	2282	106, 120, 182, 196, 211, 212[a]	Polyurethane

[a]Compounds used to calculate the signal-to-noise ratios by extracting the m/z profiles of the signal
[b]Retention time

to amplified toxicological effects. The other factor that enhances the toxicity of smaller nanoplastics is their enhanced reactive surface, which binds other hydrophobic contaminants and acts as a vector for them.

The limited information on the penetration capabilities of nanoplastics hinders the development of simulation models to predict their distribution and accumulation in crops and microorganisms realistically. Such models are key to understanding chronic and acute health risks associated with nanoplastic exposure. In the absence of nanoplastic inventories and size information, it is challenging to develop accurate models that can predict the potential health impacts of these particles over time. Elevated nanoplastic concentrations in soil can lead to more significant exposure for microorganisms and, in turn, can compromise crop vitality.

The pyro-GCMS is a potent tool for identifying polymers, but its inability to provide quantitative size and morphological information on nanoplastics complicates the exposure assessment of soil microbes and invertebrates, particularly in terms of nanoplastic effect at the molecular and cellular level. The toxicological outcomes and capability for long-term monitoring and assessment in the absence of such size and count information have high uncertainties. Also, the data generated is often incomparable as it is usually reported in terms of the concentration of polymer per unit mass rather than the number of particles per unit mass or area reported for microplastics.

The experimental toxicological data on nanoplastics is often shown to be size-dependent. The surface area to volume ratio of nanoplastics is a key determinant of their reactivity and potential to be transport vectors for contaminants of concern. Smaller nanoplastics induce higher oxidative stress and inflammation in biological systems (Zhou et al. 2023), increasing toxicity (Geng et al. 2023; Lee et al. 2024; Jinadasa et al. 2023; Uddin et al. 2023).

5.6 Conclusion

Nanoplastics are a significant concern of the Anthropocene, posing an emerging challenge to the soil and crop systems. The nanoplastics have potentially far-reaching consequences for long-term soil health, crop vitality and productivity, and food safety issues. Nanoplastic research is still in its initial stages due to technological limitations in identifying them in environmental samples. The available evidence, primarily based on experimental data, underpins the significant soil-plant interactions, negative crop growth, and food safety concerns emanating from their presence in various plant tissues. A multidisciplinary team of environmental scientists, agronomists, toxicologists, and policymakers must address this multifaceted challenge. To develop effective management and mitigation strategies for nanoplastic risks in soil and crop systems, data on the sources, behaviour in soil, and response to microbial interactions are needed.

The technical limitations of the pyro-GCMS in providing size and count (numbers) information in environmental samples hamper our ability to assess toxicity accurately. Moreover, in the absence of size information, the surface-to-volume ratio of nanoplastics, its reactive surface and toxicity are questionable. Since risk assessments typically rely on detailed information about the particle's physico-chemical characteristics, the lack of such details compromises the efficacy risk assessment. The other concerns emanate from the purity of the media, laboratory blanks, and lack of standard reference material. Most researchers have not reported background concentrations; hence, the data cannot be corrected.

To complement the information on polymer types generated by pyro-GCMS and to overcome the limitations, it is prudent to explore alternative and complementary techniques like dynamic light scattering (DLS) and nanoparticle tracking analysis (NTA) that can provide information on the nanoplastic size distribution and concentration in a sample. It will be interesting to assess how these techniques can be used to obtain a more comprehensive understanding of the characterisation of nanoplastics in environmental samples and to evaluate their realistic toxicological implications.

Furthermore, the inability to assess the size and number of nanoplastics limits the ability to establish exposure-response relationships, which are critical for determining safe levels of exposure and developing regulatory guidelines. Without this information, it is challenging to develop comprehensive risk assessments that fully account for the potential health impacts of nanoplastics. Also, the data produced is incomparable to most of the data on microplastics in published literature. It is prudent to mention that a large volume of literature on nanoplastics has inappropriately combined the information on micro-nanoplastics to the extent that some information is inaccurate and misleading; we have highlighted some of these in this perspective.

Acknowledgements The authors thank the Kuwait Institute for Scientific Research and the International Atomic Energy Agency for their support.

References

Adeel M et al (2021) Bio-interaction of nano and bulk lanthanum and ytterbium oxides in soil system: biochemical, genetic, and histopathological effects on Eisenia fetida. J Hazard Mater 415:125574

Alexander J et al (2016) Presence of microplastics and nanoplastics in food, with particular focus on seafood EFSA Panel on Contaminants in the Food Chain (CONTAM). EFSA J 14:4501

Ali I et al (2020) Advances in nanoparticles as anticancer drug delivery vector: need of this century. Curr Pharm Des 26(15):1637–1649

Alimi OS et al (2018) Microplastics and nanoplastics in aquatic environments: aggregation, deposition, and enhanced contaminant transport. Environ Sci Technol 52:1704–1724

Aralappanavar VK et al (2024) Effects of microplastics on soil microorganisms and microbial functions in nutrients and carbon cycling – a review. Sci Total Environ 924:171435

Astner AF et al (2019) Mechanical formation of micro- and nano-plastic materials for environmental studies in agricultural ecosystems. Sci Total Environ 685:1097–1106

Atugoda T et al (2023) Nanoplastic occurrence, transformation and toxicity: a review. Environ Chem Lett 21(1):363–381

Bandmann V et al (2012) Uptake of fluorescent nano beads into BY2-cells involves clathrin-dependent and clathrin-independent endocytosis. FEBS Lett 586(20):3626–3632

Bao X et al (2024) Microplastics derived from plastic mulch films and their carrier function effect on the environmental risk of pesticides. Sci Total Environ 924:171472

Bellingeri A et al (2019) Combined effects of nanoplastics and copper on the freshwater alga Raphidocelis subcapitata. Aquat Toxicol 210:179–187

Bosker T et al (2019) Microplastics accumulate on pores in seed capsule and delay germination and root growth of the terrestrial vascular plant Lepidium sativum. Chemosphere 226:774–781

Cai Z et al (2023) Biological degradation of plastics and microplastics: a recent perspective on associated mechanisms and influencing factors. Microorganisms 11(7):1661

Cao J et al (2022) Enhanced microalgal toxicity due to polystyrene nanoplastics and cadmium co-exposure: from the perspective of physiological and metabolomic profiles. J Hazard Mater 427:127937

Capozzi F et al (2018) Evidence on the effectiveness of mosses for biomonitoring of microplastics in fresh water environment. Chemosphere 205:1–7

Cavazzoli S et al (2023) Analysis of micro- and nanoplastics in wastewater treatment plants: key steps and environmental risk considerations. Environ Monit Assess 195(12):1483

Ceylan E et al (2024) Interactıon of micro-nanoplastics and heavy metals in soil systems: mechanism and implication. In: Management of micro and nano-plastics in soil and biosolids. Springer, Cham, pp 163–201

Chae Y, An Y-J (2020) Nanoplastic ingestion induces behavioral disorders in terrestrial snails: trophic transfer effects via vascular plants. Environ Sci Nano 7(3):975–983

Chen G et al (2022a) Effects of micro(nano)plastics on higher plants and the rhizosphere environment. Sci Total Environ 807(Pt 1):150841

Chen G et al (2022b) Effects of micro (nano) plastics on higher plants and the rhizosphere environment. Sci Total Environ 807:150841

Chen X et al (2022c) Physiological response and oxidative stress of grass carp (Ctenopharyngodon idellus) under single and combined toxicity of polystyrene microplastics and cadmium. Ecotoxicol Environ Saf 245:114080

Covernton GA et al (2019) Size and shape matter: a preliminary analysis of microplastic sampling technique in seawater studies with implications for ecological risk assessment. Sci Total Environ 667:124–132

Davranche M et al (2020) Nanoplastics on the coast exposed to the North Atlantic Gyre: evidence and traceability. NanoImpact 20:100262

Dong Y et al (2021) Uptake of microplastics by carrots in presence of As (III): combined toxic effects. J Hazard Mater 411:125055

Dovidat LC et al (2020) Plastic particles adsorb to the roots of freshwater vascular plant Spirodela polyrhiza but do not impair growth. Limnol Oceanograph Lett 5(1):37–45

Ferreira-Filipe DA et al (2024) Microplastics from agricultural mulch films: biodegradation and ecotoxicity in freshwater systems. Sci Total Environ 912:169287

Frehland S et al (2020) Long-term assessment of nanoplastic particle and microplastic fiber flux through a pilot wastewater treatment plant using metal-doped plastics. Water Res 182:115860

Gao Z-Y et al (2022) Single and combined toxicity of polystyrene nanoplastics and copper on Platymonas helgolandica var. tsingtaoensis: perspectives from growth inhibition, chlorophyll content and oxidative stress. Sci Total Environ 829:154571

Geng Y et al (2023) Toxicity of microplastics and nanoplastics: invisible killers of female fertility and offspring health. Front Physiol 14:1254886

Gigault J et al (2016) Marine plastic litter: the unanalyzed nano-fraction. Environ Sci Nano 3(2):346–350

Gigault J et al (2018a) Current opinion: what is a nanoplastic? Environ Pollut 235:1030

Gigault J et al (2018b) Current opinion: what is a nanoplastic? Environ Pollut 235:1030–1034

Gigault J et al (2021) Nanoplastics are neither microplastics nor engineered nanoparticles. Nat Nanotechnol 16(5):501–507

Giorgetti L et al (2020) Exploring the interaction between polystyrene nanoplastics and Allium cepa during germination: internalization in root cells, induction of toxicity and oxidative stress. Plant Physiol Biochem 149:170–177

Gupta C et al (2022) Bioaccumulation and toxicity of polystyrene nanoplastics on marine and terrestrial organisms with possible remediation strategies: a review. Environ Adv 8:100227

Hernandez LM, Yousefi N, Tufenkji N (2017) Are there nanoplastics in your personal care products? Environ Sci Technol Lett 4(7):280–285

Hurley RR, Nizzetto L (2018) Fate and occurrence of micro(nano)plastics in soils: knowledge gaps and possible risks. Curr Opin Environ Sci Health 1(6–11):6

Hussain K et al (2025) Impact of polystyrene nanoplastics on physiology, nutrient uptake, and root system architecture of aeroponically grown citrus plants. NanoImpact 37:100536

Jia L et al (2023) Microplastic stress in plants: effects on plant growth and their remediations. Front Plant Sci 14:1226484

Jiang X et al (2019) Ecotoxicity and genotoxicity of polystyrene microplastics on higher plant Vicia faba. Environ Pollut 250:831–838

Jinadasa BKKK, Uddin S, Fowler SW (2023) Microplastics (MPs) in marine food chains: is it a food safety issue? In: Ozogul F (ed) Advances in food and nutrition research. Academic Press, pp 101–140

Jones NR et al (2024) Identifying laboratory sources of microplastic and nanoplastic contamination from the air, water, and consumables. J Hazard Mater 465:133276

Khan AR et al (2024) Micro/nanoplastics: critical review of their impacts on plants, interactions with other contaminants (antibiotics, heavy metals, and polycyclic aromatic hydrocarbons), and management strategies. Sci Total Environ 912:169420

Kwon J-H et al (2017) Microplastics as a vector of hydrophobic contaminants: importance of hydrophobic additives. Integr Environ Assess Manag 13(3):494–499

Lee WS et al (2019) Bioaccumulation of polystyrene nanoplastics and their effect on the toxicity of Au ions in zebrafish embryos. Nanoscale 11(7):3173–3185

Lee J-Y et al (2024) A comprehensive review of urban microplastic pollution sources, environment and human health impacts, and regulatory efforts. Sci Total Environ 946:174297

Li P et al (2020a) Analytical methods and environmental processes of nanoplastics. J Environ Sci (China) 94:88–99

Li L et al (2020b) Effective uptake of submicrometre plastics by crop plants via a crack-entry mode. Nat Sustain 3(11):929–937

Li C et al (2021a) Quantification of Nanoplastic uptake in cucumber plants by pyrolysis gas chromatography/mass spectrometry. Environ Sci Technol Lett 8(8):633–638

Li Z et al (2021b) The distribution and impact of polystyrene nanoplastics on cucumber plants. Environ Sci Pollut Res 28(13):16042–16053

Li S et al (2021c) Polystyrene microplastics disturb the redox homeostasis, carbohydrate metabolism and phytohormone regulatory network in barley. J Hazard Mater 415:125614

Li J et al (2023) Single and combined toxicity of polystyrene nanoplastics and arsenic on submerged plant Myriophyllum verticillatum L. Plant Physiol Biochem 194:513–523

Lian J et al (2020a) Impact of polystyrene nanoplastics (PSNPs) on seed germination and seedling growth of wheat (Triticum aestivum L.). J Hazard Mater 385:121620
Lian J et al (2020b) Do polystyrene nanoplastics affect the toxicity of cadmium to wheat (Triticum aestivum L.)? Environ Pollut 263:114498
Liu L, Fokkink R, Koelmans AA (2016) Sorption of polycyclic aromatic hydrocarbons to polystyrene nanoplastic. Environ Toxicol Chem 35(7):1650–1655
Liu M et al (2018a) Microplastic and mesoplastic pollution in farmland soils in suburbs of Shanghai, China. Environ Pollut 242:855–862
Liu J et al (2018b) Polystyrene Nanoplastics-enhanced contaminant transport: role of irreversible adsorption in glassy polymeric domain. Environ Sci Technol 52(5):2677–2685
Liu Y et al (2021a) Transport and transformation of microplastics and nanoplastics in the soil environment: a critical review. Soil Use Manag 37(2):224–242
Liu Q et al (2021b) Microplastics and nanoplastics: emerging contaminants in food. J Agric Food Chem 69(36):10450–10468
Liu Y et al (2022) Uptake and translocation of nano/microplastics by rice seedlings: evidence from a hydroponic experiment. J Hazard Mater 421:126700
Ma Y et al (2016) Effects of nanoplastics and microplastics on toxicity, bioaccumulation, and environmental fate of phenanthrene in fresh water. Environ Pollut 219:166–173
Marchesan S, Prato M (2013) Nanomaterials for (Nano)medicine. ACS Med Chem Lett 4(2):147–149
Matsui K et al (2020) Identification algorithm for polymer mixtures based on Py-GC/MS and its application for microplastic analysis in environmental samples. J Anal Appl Pyrolysis 149:104834
Mclaughlin M, Pennock D (2018) Soil pollution: a hidden reality. FAO, Rome
Melo TO, Marques FA, Hansel FA (2017) Pyrolysis-gas chromatography–mass spectrometry Kováts retention index of pyrolysis products of lignocellulosic materials. J Anal Appl Pyrolysis 126:332–336
Miller ME, Hamann M, Kroon FJ (2020) Bioaccumulation and biomagnification of microplastics in marine organisms: a review and meta-analysis of current data. PLoS One 15(10):e0240792
Nath S et al (2024) Microplastics and nanoplastics in soil: sources, impacts, and solutions for soil health and environmental sustainability. J Environ Qual 53:n/a–n/a
Ng E-L et al (2018) An overview of microplastic and nanoplastic pollution in agroecosystems. Sci Total Environ 627:1377–1388
Nizzetto L, Futter M, Langaas S (2016) Are agricultural soils dumps for microplastics of urban origin? Environ Sci Technol 50(20):10777–10779
Omidoyin KC, Jho EH (2024) Environmental occurrence and ecotoxicological risks of plastic leachates in aquatic and terrestrial environments. Sci Total Environ 954:176728
Pérez-Reverón R et al (2023) Nanoplastics in the soil environment: analytical methods, occurrence, fate and ecological implications. Environ Pollut 317:120788
Petersen E et al (2023) Evaluation of bioaccumulation of nanoplastics, carbon nanotubes, fullerenes, and graphene family materials. Environ Int 173:107650
Pradel A, Catrouillet C, Gigault J (2023) The environmental fate of nanoplastics: what we know and what we need to know about aggregation. NanoImpact 29:100453
Ramanayaka S, Zhang H, Semple KT (2024) Environmental fate of microplastics and common polymer additives in non-biodegradable plastic mulch applied agricultural soils. Environ Pollut 363:125249
Rochman CM et al (2013a) Classify plastic waste as hazardous. Nature 494:169
Rochman CM et al (2013b) Ingested plastic transfers hazardous chemicals to fish and induces hepatic stress. Sci Rep 3:3263
Rochman CM, Hentschel BT, Teh SJ (2014) Long-term sorption of metals is similar among plastic types: implications for plastic debris in aquatic environments. PLoS One 9(1):e85433
Rodríguez-Hernández AG et al (2020) 3D printer waste, a new source of nanoplastic pollutants. Environ Pollut 267:115609

Roy T, Dey TK, Jamal M (2022) Microplastic/nanoplastic toxicity in plants: an imminent concern. Environ Monit Assess 195(1):27

Seo Y et al (2024) Micro- and nanoplastics in agricultural soils: assessing impacts and navigating mitigation. Sci Total Environ 931:172951

Shen M et al (2019) Recent advances in toxicological research of nanoplastics in the environment: a review. Environ Pollut 252:511–521

Shi D, Guo Z, Bedford N (2015) 9 - Nanobiological Materials. In: Shi D, Guo Z, Bedford N (eds) Nanomaterials and devices. William Andrew Publishing, Oxford, pp 215–253

Sieber G et al (2022) Microbial community shifts induced by plastic and zinc as substitutes of tire abrasion. Sci Rep 12(1):18684

Stephens B et al (2013) Ultrafine particle emissions from desktop 3D printers. Atmos Environ 79:334–339

Sun X-D et al (2020) Differentially charged nanoplastics demonstrate distinct accumulation in Arabidopsis thaliana. Nat Nanotechnol 15:755

Sun H et al (2021) Foliar uptake and leaf-to-root translocation of nanoplastics with different coating charge in maize plants. J Hazard Mater 416:125854

Sun N et al (2023) Combined toxicity of micro/nanoplastics loaded with environmental pollutants to organisms and cells: role, effects, and mechanism. Environ Int 171:107711

Tang N et al (2022) The adsorption of arsenic on micro- and nano-plastics intensifies the toxic effect on submerged macrophytes. Environ Pollut 311:119896

Taylor SE et al (2020) Polystyrene nano- and microplastic accumulation at Arabidopsis and wheat root cap cells, but no evidence for uptake into roots. Environ Sci Nano 7(7):1942–1953

Uddin S, Fowler SW, Behbehani M (2020) An assessment of microplastic inputs into the aquatic environment from wastewater streams. Mar Pollut Bull 160:111538

Uddin S et al (2022) Microplastics in Kuwait's wastewater streams. Sustainability 14(23):15817

Uddin AS, Uddin S, Fowler SW (2023) Environmental microplastics: a significant pollutant of the Anthropocene. In: Khan A, Wang C, Asiri AM (eds) Microplastic sources, fate and solution. Springer Nature Singapore, Singapore, pp 89–105

Varg JE, Bergvall C, Svanbäck R (2022) The stressful effects of microplastics associated with chromium (VI) on the microbiota of Daphnia Magna. Front Environ Sci 10:875512

Velzeboer I, Kwadijk CJAF, Koelmans AA (2014) Strong sorption of PCBs to Nanoplastics, microplastics, carbon nanotubes, and fullerenes. Environ Sci Technol 48(9):4869–4876

Wahl A et al (2021) Nanoplastic occurrence in a soil amended with plastic debris. Chemosphere 262:127784

Wahl A et al (2024) Nano(micro)plastic mobility in soil: metallic additives and Sr isotopes as potential tracers. J Hazard Mater 476:135153

Wan T et al (2019) A spectroscopic and theoretical investigation of interaction mechanisms of tetracycline and polystyrene nanospheres under different conditions. Environ Pollut 249:398–405

Wan J-K et al (2021) Influence of polystyrene microplastic and nanoplastic on copper toxicity in two freshwater microalgae. Environ Sci Pollut Res 28(25):33649–33668

Wang L et al (2020) Environmental fate, toxicity and risk management strategies of nanoplastics in the environment: current status and future perspectives. J Hazard Mater 401:123415

Wu X et al (2020) Transport of polystyrene nanoplastics in natural soils: effect of soil properties, ionic strength and cation type. Sci Total Environ 707:136065

Xiong Y et al (2020) Interfacial interaction between micro/nanoplastics and typical PPCPs and nanoplastics removal via electrosorption from an aqueous solution. Water Res 184:116100

Xu J et al (2023a) Nano- and micro-plastic transport in soil and groundwater environments: sources, behaviors, theories, and models. Sci Total Environ 904:166641

Xu G, Lin X, Yu Y (2023b) Different effects and mechanisms of polystyrene micro- and nanoplastics on the uptake of heavy metals (Cu, Zn, Pb and Cd) by lettuce (Lactuca sativa L.). Environ Pollut 316:120656

Yilimulati M et al (2021) Adsorption of ciprofloxacin to functionalized nanosized polystyrene plastic: kinetics, thermochemistry and toxicity. Sci Total Environ 750:142370

Yu J et al (2022) Heterogeneity effects of nanoplastics and lead on zebrafish intestinal cells identified by single-cell sequencing. Chemosphere 289:133133

Zhang M et al (2019) Microplastics from mulching film is a distinct habitat for bacteria in farmland soil. Sci Total Environ 688:470–478

Zhou C-Q et al (2021) Response of rice (Oryza sativa L.) roots to nanoplastic treatment at seedling stage. J Hazard Mater 401:123412

Zhou Y et al (2023) Nanoplastics induces oxidative stress and triggers lysosome-associated immune-defensive cell death in the earthworm Eisenia fetida. Environ Int 174:107899

Chapter 6
Sources of Microplastics in the Soil and Use of Laser Direct Infrared Imaging for Microplastic Detection

Saif Uddin, Montaha Behbehani, Shua'a Al-Ruwayeh, Yasmeen Al-Babtain, Maha Al-Sinan, and Ahmad Al-Shemmari

Abstract Agricultural soils have emerged as critical sinks for microplastics (MPs), with contamination driven mainly by plastic mulch films, sewage sludge, wastewater irrigation, organic fertilizers, and atmospheric deposition. These MPs, ranging from 1 μm to 5 mm, perturb the soil ecosystems by adsorbing heavy metals, organic pollutants, and antibiotic resistance genes while disrupting soil structure, microbial communities, and plant health. Despite their pervasive presence, conventional detection methods—such as FTIR and Raman spectroscopy—remain labour-intensive, making any large-scale assessments challenging. This study underscores the transformative potential of laser direct infrared (LDIR) imaging, a high-throughput technique leveraging quantum cascade lasers and rapid spectral scanning in the 975–1800 cm^{-1} range to characterize MPs by polymer type, size, and morphology with minimal pretreatment. Global data highlight alarming MP accumulation in sludge-amended soils, retaining up to 3.50 MP g^{-1} after five years, while mulched farmlands in China exhibit 1.08 MP g^{-1} after 24 years. Untreated wastewater irrigation further introduces 8–15,620 MP m^{-3} into soils, exacerbating contamination. LDIR's efficiency, demonstrated by 80–100% recovery rates for 200–500 μm particles and analysis times of ~3 hours per sample, addresses critical gaps in current protocols. Its integration with machine learning and expanding spectral libraries promises enhanced precision for sub-10 μm particles, enabling scalable monitoring. By advancing detection accuracy and standardisation, LDIR emerges as a pivotal tool for elucidating MP dynamics, guiding policy, and mitigating risks to soil health, biodiversity, and food security in the face of this escalating environmental crisis.

Keywords Soil Health · Sink · Mulch · Sewage sludge · Wastewater

S. Uddin (✉) · M. Behbehani · S. Al-Ruwayeh · Y. Al-Babtain · M. Al-Sinan · A. Al-Shemmari
Environment and Life Sciences Research Centre, Kuwait Institute for Scientific Research, Safat, Kuwait
e-mail: sdin@kisr.edu.kw

S. Uddin, M. Zaman (eds.), *Micro- and Nano-plastics in Soil and Crop Systems*, Sustainability Sciences in Asia and Africa, https://doi.org/10.1007/978-981-95-2740-3_6

6.1 Introduction

The agricultural soils are becoming a gigantic sink of microplastics (MPs) (Fan et al. 2023) due to the large-scale use of plastic mulch films (Wang et al. 2024; Ramanayaka et al. 2024; Ferreira-Filipe et al. 2024; Bao et al. 2024), organic fertilizers (Zhang et al. 2022), sewage sludge application (Ding et al. 2020; Gui et al. 2021; Baloch et al. 2023; Zaman and Uddin 2024; Uddin et al. 2024) and use of wastewater for irrigation (Uddin et al. 2020a, 2022a, 2023a; Naji et al. 2021; Lee et al. 2024). The compositional diversity of the polymers is another complex issue as more and more polymers are developed to meet the ever-increasing demand in terms of durability, lightweight, affordability, and water and weather resistance. The MPs possess traits that magnify their ecological threat: minuscule size between 1–5000 μm, vast surface area, and hydrophobic properties (Moller et al. 1989; Uddin et al. 2020b, 2022b; Jinadasa et al. 2023). These characteristics enable them to adsorb toxic pollutants, including heavy metals (Yongming et al. 2018; Feng et al. 2023), organic contaminants (Xia et al. 2023), and antibiotic resistance genes (Uddin et al. 2023b), acting as vectors for environmental contamination. Their mobility via wind and water further amplifies their reach, allowing them to infiltrate soils through agricultural practices, wastewater irrigation, flooding or atmospheric deposition. Once embedded in terrestrial ecosystems, microplastics disrupt soil integrity by altering porosity, pH, and nutrient cycles (Wang et al. 2022), and destabilizing microbial communities critical for soil fertility (Liu et al. 2019a). More alarmingly, microplastics impair plant health by obstructing seed germination, stunting root and shoot growth, and infiltrating plant tissues—even translocating nanoplastics (as small as 50 nm) into cellular structures (Jia et al. 2023; Xu et al. 2023a; Sun et al. 2023). This bioaccumulation threatens food chains, ultimately jeopardizing human health. Compounding these risks, microplastics synergize with adsorbed pollutants, exacerbating soil degradation and creating persistent ecological hazards (Liu et al. 2019b).

Despite the growing volume of literature, with currently over 900 research articles on soil microplastics—critical gaps persist. Most studies have used methods like stereo microscopy, fluorescence microscopy, ATR-Fourier transform infrared (FTIR) spectroscopy (Jia et al. 2022), RAMAN spectroscopy (Uddin et al. 2021), and scanning electron microscopy (Mariano et al. 2021)for MP identification, often overlooking advanced detection methods such as laser direct infrared imaging or matrix-assisted laser desorption ionization (MALDI-TOF), that can give information on size, shape, and polymer type. Understanding the sources, distribution, and impacts of soil microplastics is a pressing environmental issue. Refining MP detection protocols and unravelling their ecological consequences can safeguard soil health, which is the foundation of food security, biodiversity, and planetary resilience against this pervasive threat. The existing techniques for isolating microplastics from complex soil matrices remain inconsistent and underdeveloped.

Despite efforts made towards developing a standardised methodology for the extraction and detection of MPs, critical gaps exist and improvement is needed. A standard has been developed for soil sampling, i.e. ISO 18400-102:2017 "Selection and application of sampling techniques" (ISO 2017). However, due to differences in processing techniques for MPs, a unified approach is yet to be agreed upon to assess global contamination patterns and inform mitigation strategies.

Most of the popularly applied techniques are quite time-consuming and labour-intensive, which prompts us to suggest using the laser direct infrared (LDIR) imaging technique (Ghanadi et al. 2024) as a plausible alternative, as it is both fast and reliable.

Laser direct infrared imaging system employs an infrared (IR) microscope spectrometer linked to a high-speed scanning mechanism, utilizing a rapidly tunable quantum cascade laser (QCL) as the light source. LDIR's imaging system can analyse particle count, size, and morphology, while the spectrometer identifies polymer types (Scircle et al. 2020). The QCL is semiconductor based and generates light through electron tunnelling across engineered quantum wells. Here, photon wavelength is controlled by the thickness and arrangement of semiconductor layers rather than material composition (Stock et al. 2019). This design enables rapid wavelength tuning, surpassing traditional monochromator-based systems in speed and precision. The LDIR combines the QCL with a thermoelectrically cooled mercury cadmium telluride (MCT) detector and dynamic optics, operating in two modes, i.e., spatial scanning, where the laser remains at a fixed wavelength while the objective scans the sample. Each pixel is analysed in ~40 μs, enabling rapid large-area mapping. The other mode is the spectral scanning, where the laser sweeps across wavelengths at a fixed sample position, capturing a full mid-infrared spectrum (1800–975 cm^{-1}) in under 1 second. For the analysis of microplastic, both modes were utilized. Initial spatial scans (~145–1100 mm^2 area) identified particles, recording their size, shape, and position. Subsequent spectral scans at each particle's location generated mid-IR spectra, cross-referenced with a custom spectral library derived from published databases (Primpke et al. 2018). More and more spectra are added to this database, further enhancing the capability and utility of the process to identify polymers. Particles under 30 μm triggered automated refocusing, extending per-particle analysis to 8 seconds.

6.2 Sources of Microplastics in Soil

The soil is a sink for a gigantic amount of microplastics that emanate from different sources not limited to plastic degradation in waste dumps, sewage sludge application to soils, plastic mulch films used in agricultural farms, wastewater used for irrigation, organic fertilizers, and atmospheric deposition.

6.2.1 Sewage Sludge Application

Several studies have highlighted (Uddin et al. 2020a, 2022a; Naji et al. 2021; Gies et al. 2018; Li et al. 2018a; Lares et al. 2018; Talvitie et al. 2015, 2017; Mintenig et al. 2017; Magni et al. 2019; Magnusson and Norén 2014; Magnusson et al. 2016) the increased presence of microplastics in sludge. In our previous study, we computed that 88–99.9% of MPs are retained in sludge (Uddin et al. 2020a); in another study, a comparable estimate was given by Sun et al. (2019), indicating 98.3% retention of MPs in sludge. In a major wastewater treatment plant (WWTP) in Vancouver, British Colombia, the primary sludge has 14.9 ± 6.3 MP g^{-1} and the secondary sludge has 4.4 ± 2.8 MP g^{-1} (Gies et al. 2018). In a study on 28 WWTPs in China, 22.7 ± 12.1 MP g^{-1} was reported in sludge (Li et al. 2018a), whereas in sludge from Mikkeli WWTP, Finland, the load was 27.3 ± 4.7 MP g^{-1} and in Kenkaveronniemi WWTP,the microplastics in activated sludge were 23.0 ± 4.2 MP g^{-1} and in digested sludge 170.9 ± 28.7 MP g^{-1} (Lares et al. 2018). Microplastics in dry sludge from Viikinmaki WWTP, Finland, are 186 ± 26.0 MP g^{-1} (Talvitie et al. 2017), while in Oldenbury WWTP, Germany, it was 1 MP g^{-1}; whereas in Scharrel WWTP, Germany, the concentration was 24 MP g^{-1} (Mintenig et al. 2017). Magni et al. (2019) found 59.5 MP g^{-1} in sludge from a WWTP in northern Italy. The MPs in sludge reported from Heenvliet, Westpoort, and Amsterdam West WWTPs, Netherlands, were 660 ± 410, 510 and 760 MP kg^{-1} ww (Leslie et al. 2017), respectively. Microplastics in sludge in sewage from Langeviksverket WWTP, Sweden, were 16.7 ± 1.96 MP g^{-1} (Magnusson and Norén 2014). These data provide an idea of the magnanimous quantity of MPs that exist in sludge, and their application to agricultural soils can result in accumulation. Annual microplastic input from sludge is estimated at 0.63–4.30 million tons in Europe and 0.44–3.00 million tons in North America (Ng et al. 2018). In China, this figure reaches approximately 1.56×10^{14} particles per year (Li et al. 2018b).

The recurrent sludge application results in microplastic build-up in the soil (Corradini et al. 2021). After five years of sludge application, microplastic densities can average 3.50 MP g^{-1} (Corradini et al. 2019), significantly higher than soils free of sludge application (Zhang et al. 2020). The sludge-applied soil acts as a sink for MPs, changing its physico-chemical characteristics like bulk density, porosity, and permeability. It also affects the level of metals and organic pollutants in the soil. Some researchers have suggested the clean-up of microplastics from the soil by incineration (Xu et al. 2023b), which is quite impractical. Another feasible alternative is hydrothermal treatment of sludge, heating it to 200–300 °C. However, this is cost-intensive. More research is needed to assess the ecological implications of adding sludge, while a study revealed that crops like wheat and lettuce can absorb microplastics from sludge-amended soils, posing food safety risk; the mechanism of uptake is still not clearly elucidated. The evidence for MPs affecting soil properties is equally strong to the data that suggest minimal negative effects and considerable benefits of adding sludge that provides the nutrient input and organic content to the soil. A review emphasized the lack of global standards for microplastics in sludge, with only Germany and Switzerland setting provisional guidelines (de Ruijter et al. 2020); there is an eminent need for an environmental and social impact assessment of sludge addition to agricultural soils (Zaman and Uddin, 2024).

6.2.2 Wastewater Irrigation

The omnipresence and prevalence of MPs in wastewater are well documented. Although the WWTPs are known to trap MPs in sludge and reduce the concentration in effluent, a considerable amount of MP input takes place. Moreover, about 50% of the wastewater globally remains untreated, and a good part of it is still utilized for agriculture. The MP concentration in effluent water from Australia was 934 MP m^{-3} (Ziajahromi et al. 2017), whereas in British Columbia, Canada, the average influent and effluent concentrations were 31,100 and 1550 MP m^{-3} (Gies et al. 2018), respectively. The average influent and effluent concentrations of WWTPs in Denmark are 7827 and 120 MP m^{-3} (Simon et al. 2018), respectively. The data from the WWTP in Mikkeli, Finland, have influent and effluent concentrations of 57,400 MP m^{-3} and 700 MP m^{-3}, respectively (Lares et al. 2018). In another study in Finland, the influent MP concentration at Viikinmaki WWTP was 430,000 MP m^{-3} and the influent and effluent MP concentrations from Suomenoja WWTP were 380,000 MP m^{-3} and 4100 MP m^{-3}, respectively (Talvitie et al. 2015). The average influent and effluent concentrations at Seine-Centre WWTP, France, were 293,000 MP m^{-3} and 35,000 MP m^{-3} (Dris et al. 2015), respectively, whereas much higher effluent concentrations of 15,620 MP m^{-3} were reported from Jade Bight WWTP, Germany (Dubaish and Liebezeit 2013). Data from the WWTP in Italy reported that the influent MP concentration was 2000 MP m^{-3} and effluent concentration was 300 MP m^{-3} (Magni et al. 2019). From Westport WWTP, Netherlands, the average influent and effluent concentrations were 276,000 MP m^{-3} and 58,380 MP m^{-3}, respectively (Leslie et al. 2017). Murphy et al. (2016) reported an influent concentration of 15,700 MP m^{-3} and an effluent concentration of 4120 MP m^{-3} from Glasgow WWTP, UK. An influent MP concentration of 12,430 MP m^{-3} and an effluent concentration of 4120 MP m^{-3} were reported from Cabezo Beaza WWTP, Spain (Bayo et al. 2020). The MP concentration in Langeviksverket WWTP, Sweden, in the influent was 15,100 MP m^{-3} and in the effluent was 8.25 MP m^{-3} (Magnusson and Norén 2014). Microplastics in influent and effluent streams of the Seyhan WWTP, Turkiye, were 25,000 MP m^{-3} and 5555 MP m^{-3}, respectively (Gündoğdu et al. 2018). In the Detroit WWTP, USA, the average influent MP concentration was 510,600 MP m^{-3} and the effluent concentration was 81,144 MP m^{-3} (Michielssen et al. 2016). The global data indicate that effluent water has a significant MP load, and given the water shortages, the untreated and partially treated wastewater is used for irrigation, leading to MP build-up in soil (WHO 2013).

6.2.3 Mulch Films

Plastic mulch films are extensively utilized in modern agriculture to enhance crop productivity, soil moisture retention, and weed suppression (Samphire et al. 2023). Global plastic production reached 425 million tons in 2018, with China contributing 30% of the total production (Sa'adu and Farsang 2023). Despite the agronomic benefits of mulch films usage, these films degrade quickly and are seldom recovered and less than 60% is recycled (Zhaorong et al. 2020) leading to significant environmental concerns, particularly soil microplastic pollution. The most commonly used mulch films are polyethene (PE), known to degrade due to physical degradation, including tillage, harvesting, and weathering from UV exposure, mechanical stress, and microbial activity, leading to the generation of secondary microplastics (Steinmetz et al. 2016; Zhang et al. 2021). Residual plastic fragments from these mulch films persist in soils for decades.

Plastic mulch films improve crop yields by regulating soil temperature, conserving water, and suppressing weeds. These benefits are critical in both arid and cold regions and for high-value crops, driving widespread adoption (Kasirajan and Ngouajio 2012). There is equally compelling evidence highlighting the risks of microplastics from mulch altering soil structure, including reduced soil porosity, altered microbial communities and activities, altered nutrient cycling, and potential uptake of micro- and nanoplastics by crops (Shafea et al. 2023; Rillig et al. 2019; Rillig and Lehmann 2020). Microplastics may also adsorb toxic chemicals, posing risks to soil organisms and potentially entering the food chain (Hurley and Nizzetto 2018; Crossman et al. 2020; Kokalj et al. 2024).

Studies have demonstrated that soils under plastic mulch exhibit significantly higher microplastic concentrations than uncovered soils. Near Hangzhou Bay, China, mulched soils contained up to 1.08 ± 0.347 MP g^{-1} after 24 years of use, compared to 0.803 ± 0.493 MP g^{-1} after 5 years (Huang et al. 2020) underpinning the hypothesis that MP concentration builds up in soil over time, and soil is a sink for MPs. This was further supported by the report that the duration of use directly correlates with microplastic levels, indicating progressive fragmentation of films into <5 mm particles (Zhou et al. 2020). However, there is variation in MP concentrations within the region and globally due to differing agricultural practices and recovery rates. In Mumbai, India, MPs are reported to vary between 0.012–0.315 MP g^{-1} in mulched farmlands across rural to metropolitan areas; in another review, the paucity of data on MPs in agricultural soils in India was highlighted (Mahesh et al. 2022; Anand and Sheel 2025). While the MP concentration in soils in northern Germany was 0.0043 ± 0.012 MP g^{-1} (Harms et al. 2021) and in Tunisia, 0.880 MP g^{-1} (Uzamurera et al. 2023).

Plastic mulch, widely used in agriculture for weed suppression, moisture retention, and soil temperature regulation, has been estimated to contribute 10–15% of total MPs in agricultural soils globally, with higher contributions in plastic-intensive systems (Bläsing and Amelung 2018). China accounts for about 60% of global plastic mulch use. Fields in Xinjiang showed 200–400 kg ha^{-1} of residual plastic after 10–30 years of use (Liu et al. 2014). Qi et al. (2020) quantified 50–300 MP kg^{-1} mainly PE, in agricultural soil that was subjected to long-term plastic mulching in China. In Europe, PE mulch fragments were detected in 90% of sampled agricultural soils (Piehl et al. 2018). To address the issue of MP emanating from PE plastic mulch films, biodegradable alternatives like polylactic acid (PLA) have been suggested (Qi et al. 2020). However, PLA also degrades into MPs, albeit faster than PE (Huang et al. 2020). Research into cost-effective, fully degradable films is quite important, though current alternatives face challenges in durability and incomplete degradation; hence, their long-term ecological effects remain uncertain. The issue is further exacerbated, as most countries lack regulations for plastic mulch recovery, enhanced recycling infrastructure, incentives for farmers to remove post-harvest films, and incentivise the use of biodegradable alternatives.

6.2.4 *Organic Fertilizer*

The global shift towards sustainable agriculture has increased the utilisation of organic fertilisers. The organic fertiliser is prepared from animal manures, plant residues, and composted waste (Okoffo et al. 2021) and has become a cornerstone of eco-friendly farming. While organic fertilisers enrich soils with essential nutrients, reducing reliance on synthetic fertilisers, an insidious threat accompanies their application: the inadvertent introduction of MPs into agricultural soils. Emerging research underscores that organic fertilisers, often perceived as a "green" alternative, are a significant yet underrecognized pathway of MP introduction into the soil, with possible negative implications for ecosystem health and food security.

The MPs infiltrate into the food chain via organic fertiliser usage through absorption by plant roots and entering the food chain when consumed by livestock (Zhang et al. 2024). Analyses of organic fertilisers reveal alarming microplastic concentrations (Weithmann et al. 2018). In Bonn, Germany, plastic fragments larger than 0.50 mm were detected at levels of 2.38–180 mg kg^{-1}, while another study documented 0.014–0.895 particles g^{-1} of microplastics >1 mm in German fertilisers (Bläsing and Amelung 2018; Weithmann et al. 2018), despite stringent regulations in Germany that limit plastic to 0.1 wt% and ban MP <2.00 mm; microplastics persist due to gaps in waste processing. Similarly, Australia's National Association for Sustainable Agriculture permits ≤0.5 wt% hard plastics and ≤0.05 wt% light plastics in organic fertilisers, yet there are no guidelines set so far. The MPs in organic waste in Australia range between 1500 and 16,000 MP kg^{-1} dw, with polyethene, polypropylene, and polyester as the main polymer types in a size range of 25–500 μm (Lu et al. 2025). China is probably the largest consumer of organic fertilisers, with an annual MP contribution from organic fertiliser input to soils estimated at 52.4 tons (Zhang et al. 2022). In another study, the average MP abundance in animal-derived organic fertiliser was 10692.6 ± 3353.3 MP kg^{-1}, in organic fertiliser derived from plant and animal material, it was 7914.2 ± 4444.3 MP kg^{-1}, whereas in plant-derived organic compost, the concentration was 7539.5 ± 2779.9 MP kg^{-1} (Xu et al. 2024). The MP in farm compost was highest, with a concentration of 24344.1 ± 2697.2 MP kg^{-1} much higher than the commercial organic fertilisers where the MP ranged between 4266.7 ± 2274.5 and 18028.6 ± 2000.0 MP kg^{-1} (Xu et al. 2024). Smaller particles (<0.50 mm), which evade standard filtration methods, pose heightened risks due to their mobility and potential to disrupt soil microbiota and plant health (Fan et al. 2023; Aralappanavar et al. 2024). This trend is not isolated, as ~54% of MPs are <1.0 mm, 35.2% in the size range of 1–3 mm, ~7.8% in the size range of 3–5 mm, and ~3% >5 mm (Xu et al. 2024).

The presence of MPs in organic fertilisers underscores a paradox of sustainable agriculture: practices designed to heal ecosystems may inadvertently harm them. While organic fertilisers remain vital for soil health, their contamination highlights systemic flaws in waste management and regulatory oversight. Only through proactive measures can we safeguard soils from becoming a sink for MPs, ensuring that the benefits of organic farming are not eclipsed by its long-term liability and possibly compromised ecosystem services.

6.2.5 Atmospheric Deposition

A significant amount of literature has suggested the omnipresence of MPs in aerosol (Habibi et al. 2022) and their potential for long-range transport (Allen et al. 2019). The MPs in dust and aerosols are deposited either via wet and dry atmospheric deposition or even with snow (Dris et al. 2015, 2016, 2017, 2018; Allen et al. 2019; Abbasi et al. 2019; Akanyange et al. 2021; Gasperi et al. 2018; Liao et al. 2021; Prata et al. 2020; Soltani et al. 2021; Wright et al. 2019, 2021). MPs in the air are released from wear and tear of clothing material (Napper and Thompson 2016; Henry et al. 2019; De Falco et al. 2019), by washing and drying (Browne et al. 2011), erosion of synthetic rubber tyres, deterioration of house furniture, emissions from the synthetic textile industry, emissions from vinyl chloride (VC) and polyvinyl chloride (PVC) industries, and contamination from city dust (Dris et al. 2015, 2016, 2017, 2018). MP concentrations vary dramatically across regions with 0.3–1.5 MP m^{-3} in outdoor Paris (Dris et al. 2017) and 0.02–154 × 10^3 MP L^{-1} in snow in the Arctic (Allen et al. 2019). The MPs in China ranged between 175 and 602 particles m^{-2} day^{-1} (Cai et al. 2017; Zhou et al. 2017). In another report, 51 MP m^{-3} was reported in the mountains of Wenzhou, China (Liao et al. 2021), 275 MP m^{-2} day^{-1} in Hamburg (Klein and Fischer 2019), whereas from Indonesia (132.75–174.97 MP m^{-3}), the West Pacific Ocean (0–1.37 MP m^{-3}), Denmark (1.67–16.2 MP m^{-3}), Ireland (0–12 MP m^{-3}), Brazil (0–24 MP m^{-3}), California (0.6–5.6 MP m^{-3}), in Nottingham 2.9 MP m^{-2} day^{-1} and Central London 712 ± 162 MP m^{-2} day^{-1} were reported (Syafei et al. 2019; Liu et al. 2019c; Vianello et al.

2019; Roblin et al. 2020; Amato-Lourenço et al. 2021; Stanton et al. 2019; Wright et al. 2020). The dry and wet atmospheric deposition in Vietnam ranged between 71 and 917 MPs m^{-2} day^{-1} (Truong et al. 2021). From the available information, it is clear that atmospheric deposition might contribute MPs to soils; however, the use of inconsistent methodologies in sampling, processing, and reporting (e.g. MP m^{-2} day^{-1} vs. MP m^{-3}) hinder any reliable assessments. The capacity for long-range transport, coupled with variable sizes and concentrations, underscores the urgent need for harmonised research to quantify risks and inform mitigation strategies for MP management.

6.3 Laser Direct Infrared (LDIR) Imaging

Considering the complexity of the agricultural soil matrix, which is rich in organic matter and often laden with plant and animal debris, the spatial variation of soil types makes the extraction of MPs a complex task. The LDIR has emerged as a powerful tool for characterising microplastics in environmental samples, enabling precise identification of polymer composition, morphology, and particle dimensions. Unlike conventional approaches, LDIR eliminates the need for chemical pre-treatment, allowing rapid, non-destructive analysis that maintains the structural and chemical integrity of samples. Its high sensitivity facilitates the detection of trace microplastic levels through infrared spectral evaluation (López-Rosales et al. 2024). The automated workflow of LDIR captures mid-infrared spectra from individual particles, cross-referencing them against a polymer database for particles spanning 20–500 μm (Zhang et al. 2018; Müller et al. 2020). Using the precise and sensitive LDIR technique, an efficient MP identification, concentration, and size distribution in sediment samples is reported, which to a reasonable extent is similar to soil. The advantage of the LDIR is that instead of processing large sediment samples, comparable results are obtained with smaller sample sizes of 2–5 g (Ghanadi et al. 2024). There have been significant efforts put into developing and improving the LDIR protocol; Whiting et al. (2022) optimized LDIR protocols using polycarbonate filters (8 μm porosity), achieving 80–100% recovery rates for 200–500 μm particles across environmental matrixes (water, sediment, and biota). Their method reduced the analysis time to ~3 hr per sample, a marked improvement over traditional techniques used in agrosystems (Ng et al. 2018; Liu et al. 2022). In another study, the enzymatic-oxidative digestion coupled with LDIR showed a near-complete recovery for >500 μm particles and 75% for 10 μm fibres in GI tracts of fish (Ng et al. 2018; López-Rosales et al. 2024) suggesting its applicability in high organic matrixes. Hildebrandt et al. (2022) integrated microwave-assisted digestion with LDIR, identifying 20 polymers (predominantly PET and PU) in diverse samples, with 93% of particles <100 μm.

Several studies have effectively used LDIR for environmental monitoring and assessments in China; studies revealed microplastic concentrations of 8–132 MP

m^{-3}, primarily >100 μm (Fan et al. 2022). Studies have demonstrated that the digestion of organic matter in a sample is a crucial step for MP determination by LDIR, as incomplete digestion can lead to false positives (low degree of correlation with the spectral library of polymers). LDIR was effectively used for MPs in lake sediments in China; the identification was reported as extremely rapid at 7.2 sec per particle (Cheng et al. 2022). Two agricultural areas in China were assessed for MPs using LDIR with concentrations of 380–3786 MP kg^{-1} and phthalate esters 0.30–1.58 mg kg^{-1} in soils, noting higher microplastic loads in greenhouse sites (Li et al. 2021).

6.4 Discussion and Conclusion

Over the past decade, significant technological and methodological progress has been made in detecting and quantifying MPs in environmental matrices. Each technique—whether spectroscopy, chromatography, or imaging—offers distinct advantages and constraints, necessitating careful selection based on research goals. Several workers have put detailed accounts of the efficacies of the various techniques used for MPs (Primpke et al. 2020; Isaac Chandran and Veerasingam 2025). It is quite obvious that MP analyses are time-consuming, and it becomes quite challenging if there are a large number of possible plastic particles in an environmental matrix like sludge, sludge-applied soil, or plastic mulch-applied soils.

For environmental surveillance, a high-throughput technique is required, whereas studies on polymer toxicity or bioaccumulation pathways demand higher spectral resolution and sensitivity. To keep this balance, LDIR chemical imaging has evolved as a promising technique for MP analysis. By employing a quantum cascade laser (QCL), LDIR captures infrared spectra in as little as 1 second per particle, enabling rapid mapping of thousands of particles across large sample areas. This efficiency surpasses traditional methods like Raman spectroscopy and Fourier-transform infrared (FTIR) spectroscopy, which often require minutes per measurement. Integrated visible cameras in LDIR systems further enhance detection accuracy, resolving particles down to 10 μm—critical for identifying smaller MPs prevalent in environmental samples. The spectral range of LDIRs currently covers 975–1800 cm^{-1} and limits its ability to identify certain polymers, which exhibit characteristic peaks outside this window. Despite this, its non-destructive nature, minimal sample preparation, and high sensitivity make LDIR indispensable for MP monitoring in soil. Automation, combined with machine learning algorithms for spectral matching, could streamline high-throughput screening, enabling global MP monitoring. Innovations in QCL power and detector sensitivity may further enhance spatial resolution, allowing the detection of sub-10 μm particles—critical for understanding ecological impacts.

References

Abbasi S et al (2019) Distribution and potential health impacts of microplastics and microrubbers in air and street dusts from Asaluyeh County, Iran. Environ Pollut 244:153–164

Akanyange SN et al (2021) Does microplastic really represent a threat? A review of the atmospheric contamination sources and potential impacts. Sci Total Environ 777:146020

Allen S et al (2019) Atmospheric transport and deposition of microplastics in a remote mountain catchment. Nat Geosci 12:339–344

Amato-Lourenço LF et al (2021) Presence of airborne microplastics in human lung tissue. J Hazard Mater 416:126124

Anand V, Sheel R (2025) Microplastic contamination in agricultural soils across India : a systematic review. Int J Environ Climate Change 15(2):486–503

Aralappanavar VK et al (2024) Effects of microplastics on soil microorganisms and microbial functions in nutrients and carbon cycling – a review. Sci Total Environ 924:171435

Baloch MYJ et al (2023) Utilization of sewage sludge to manage saline–alkali soil and increase crop production: is it safe or not? Environ Technol Innov 32:103266

Bao X et al (2024) Microplastics derived from plastic mulch films and their carrier function effect on the environmental risk of pesticides. Sci Total Environ 924:171472

Bayo J, Olmos S, López-Castellanos J (2020) Microplastics in an urban wastewater treatment plant: the influence of physicochemical parameters and environmental factors. Chemosphere 238:124593

Bläsing M, Amelung W (2018) Plastics in soil: analytical methods and possible sources. Sci Total Environ 612:422–435

Browne MA et al (2011) Accumulation of microplastic on shorelines worldwide: sources and sinks. Environ Sci Technol 45(21):9175

Cai L et al (2017) Characteristic of microplastics in the atmospheric fallout from Dongguan city, China: preliminary research and first evidence. Environ Sci Pollut Res 24:24928

Cheng Y-L et al (2022) Characterization of microplastics in sediment using stereomicroscopy and laser direct infrared (LDIR) spectroscopy. Gondwana Res 108:22–30

Corradini F et al (2019) Evidence of microplastic accumulation in agricultural soils from sewage sludge disposal. Sci Total Environ 671:411–420

Corradini F et al (2021) Microplastics occurrence and frequency in soils under different land uses on a regional scale. Sci Total Environ 752:141917

Crossman J et al (2020) Transfer and transport of microplastics from biosolids to agricultural soils and the wider environment. Sci Total Environ 724:138334

De Falco F et al (2019) The contribution of washing processes of synthetic clothes to microplastic pollution. Sci Rep 9(1):6633

de Ruijter VN et al (2020) Quality criteria for microplastic effect studies in the context of risk assessment: a critical review. Environ Sci Technol 54(19):11692–11705

Ding L et al (2020) The occurrence and distribution characteristics of microplastics in the agricultural soils of Shaanxi Province, in north-western China. Sci Total Environ 720:137525

Dris R et al (2015) Microplastic contamination in an urban area: a case study in Greater Paris. Environ Chem 12:592

Dris R et al (2016) Synthetic fibers in atmospheric fallout: a source of microplastics in the environment? Mar Pollut Bull 104(1–2):290–293

Dris R et al (2017) A first overview of textile fibers, including microplastics, in indoor and outdoor environments. Environ Pollut 221:453–458

Dris R, Gasperi J, Tassin B (2018) Sources and fate of microplastics in urban areas: a focus on Paris Megacity. In: Wagner M, Lambert S (eds) Freshwater microplastics : emerging environmental contaminants? Springer International Publishing, Cham, pp 69–83

Dubaish F, Liebezeit G (2013) Suspended microplastics and black carbon particles in the Jade System, Southern North Sea. Water Air Soil Pollut 224(2):1352

Fan Y et al (2022) Spatiotemporal dynamics of microplastics in an urban river network area. Water Res 212:118116

Fan W et al (2023) Sources and identification of microplastics in soils. Soil Environ Health 1(2):100019

Feng J et al (2023) Assessing metal contamination and speciation in sewage sludge: implications for soil application and environmental risk. Rev Environ Sci Biotechnol 22(4):1037–1058

Ferreira-Filipe DA et al (2024) Microplastics from agricultural mulch films: biodegradation and ecotoxicity in freshwater systems. Sci Total Environ 912:169287

Gasperi J et al (2018) Microplastics in air: are we breathing it in? Curr Opin Environ Sci Health 1:1–5

Ghanadi M et al (2024) Quantification and characterization of microplastics in coastal environments: insights from laser direct infrared imaging. Sci Total Environ 912:168835

Gies EA et al (2018) Retention of microplastics in a major secondary wastewater treatment plant in Vancouver, Canada. Mar Pollut Bull 133:553–561

Gui J et al (2021) Microplastics in composting of rural domestic waste: abundance, characteristics, and release from the surface of macroplastics. Environ Pollut 274:116553

Gündoğdu S et al (2018) Microplastics in municipal wastewater treatment plants in Turkey: a comparison of the influent and secondary effluent concentrations. Environ Monit Assess 190(11):626

Habibi N et al (2022) Microplastics in the atmosphere: a review. J Environ Expo Assess 1(1):6

Harms IK et al (2021) Amount, distribution and composition of large microplastics in typical agricultural soils in Northern Germany. Sci Total Environ 758:143615

Henry B, Laitala K, Klepp IG (2019) Microfibres from apparel and home textiles: prospects for including microplastics in environmental sustainability assessment. Sci Total Environ 652:483–494

Hildebrandt L et al (2022) Spatial distribution of microplastics in the tropical Indian Ocean based on laser direct infrared imaging and microwave-assisted matrix digestion. Environ Pollut 307:119547

Huang Y et al (2020) Agricultural plastic mulching as a source of microplastics in the terrestrial environment. Environ Pollut 260:114096

Hurley RR, Nizzetto L (2018) Fate and occurrence of micro(nano)plastics in soils: knowledge gaps and possible risks. Curr Opin Environ Sci Health 1(6–11):6

Isaac Chandran PJ, Veerasingam S (2025) Laser direct infrared spectroscopy: a cutting-edge approach to microplastic detection in environmental samples. Talanta 284:127284

ISO (2017) ISO 18400-102 soil quality — sampling — part 102: selection and application of sampling technique. In: First edition 2017–01. I.S. Organization, Switzerland

Jia W et al (2022) Automated identification and quantification of invisible microplastics in agricultural soils. Sci Total Environ 844:156853

Jia L et al (2023) Microplastic stress in plants: effects on plant growth and their remediations. Front Plant Sci 14:1226484

Jinadasa BKKK, Uddin S, Fowler SW (2023) Microplastics (MPs) in marine food chains: is it a food safety issue? In: Ozogul F (ed) Advances in food and nutrition research. Academic Press, pp 101–140

Kasirajan S, Ngouajio M (2012) Polyethylene and biodegradable mulches for agricultural applications: a review. Agron Sustain Dev 32:501

Klein M, Fischer EK (2019) Microplastic abundance in atmospheric deposition within the Metropolitan area of Hamburg, Germany. Sci Total Environ 685:96

Kokalj AJ et al (2024) Strategy towards producing relevant and reliable data for the hazard assessment of micro- and nanoplastics in agricultural soils. TrAC Trends Anal Chem 172:117567

Lares M et al (2018) Occurrence, identification and removal of microplastic particles and fibers in conventional activated sludge process and advanced MBR technology. Water Res 133:236–246

Lee J-Y et al (2024) A comprehensive review of urban microplastic pollution sources, environment and human health impacts, and regulatory efforts. Sci Total Environ 946:174297

Leslie HA et al (2017) Microplastics en route: field measurements in the Dutch river delta and Amsterdam canals, wastewater treatment plants, North Sea sediments and biota. Environ Int 101:133–142

Li X et al (2018a) Microplastics in sewage sludge from the wastewater treatment plants in China. Water Res 142:75–85

Li J, Liu H, Paul Chen J (2018b) Microplastics in freshwater systems: a review on occurrence, environmental effects, and methods for microplastics detection. Water Res 137:362–374

Li Q et al (2021) Are microplastics correlated to phthalates in facility agriculture soil? J Hazard Mater 412:125164

Liao Z et al (2021) Airborne microplastics in indoor and outdoor environments of a coastal city in Eastern China. J Hazard Mater 417:126007

Liu EK, He WQ, Yan CR (2014) 'White revolution' to 'white pollution'—agricultural plastic film mulch in China. Environ Res Lett 9(9):091001

Liu SL et al (2019a) Pollution characteristics of microplastics in migratory bird habitats located within Poyang Lake Wetlands. Huan Jing Ke Xue 40(6):2639–2646

Liu K et al (2019b) Accurate quantification and transport estimation of suspended atmospheric microplastics in megacities: implications for human health. Environ Int 132:105127

Liu K et al (2019c) Consistent transport of terrestrial microplastics to the ocean through atmosphere. Environ Sci Technol 53(18):10612–10619

Liu N et al (2022) Characterization of microplastics in the septic tank via laser direct infrared spectroscopy. Water Res 226:119293

López-Rosales A et al (2024) A reliable method to determine airborne microplastics using quantum cascade laser infrared spectrometry. Sci Total Environ 913:169678

Lu H-C et al (2025) Microplastic in Australian processed organics: abundance, characteristics and potential transport to soil ecosystem. J Environ Manag 375:124359

Magni S et al (2019) The fate of microplastics in an Italian wastewater treatment plant. Sci Total Environ 652:602–610

Magnusson K, Norén F (2014) Screening of microplastic particles in and down-stream a wastewater treatment plant. In: IVL report C, p 22

Magnusson K et al (2016) Swedish sources and pathways for microplastics to the marine environment. Report C183. Swedish Environmental Research Institute, Stockholm

Mahesh P, Chauhan A, Warrier A (2022) Plastic mulching; microplastics in agricultural soil

Mariano S et al (2021) Micro and nanoplastics identification: classic methods and innovative detection techniques. Front Toxicol 3:636640

Michielssen MR et al (2016) Fate of microplastics and other small anthropogenic litter (SAL) in wastewater treatment plants depends on unit processes employed. Environ Sci WaterRes Technol 2(1064):1064

Mintenig SM et al (2017) Identification of microplastic in effluents of waste water treatment plants using focal plane array-based micro-Fourier-transform infrared imaging. Water Res 108:365–372

Moller L et al (1989) The air pollutant 2-nitrofluorene as initiator and promoter in a liver model for chemical carcinogenesis. Carcinogenesis 10(3):435–440

Müller A et al (2020) The pollution conveyed by urban runoff: a review of sources. Sci Total Environ 709:136125

Murphy F et al (2016) Wastewater Treatment Works (WwTW) as a source of microplastics in the aquatic environment. Environ Sci Technol 50(11):5800

Naji A et al (2021) Microplastics in wastewater outlets of Bandar Abbas city (Iran): a potential point source of microplastics into the Persian Gulf. Chemosphere 262:128039

Napper IE, Thompson RC (2016) Release of synthetic microplastic plastic fibres from domestic washing machines: effects of fabric type and washing conditions. Mar Pollut Bull 112(1):39–45

Ng E-L et al (2018) An overview of microplastic and nanoplastic pollution in agroecosystems. Sci Total Environ 627:1377–1388

Okoffo ED et al (2021) Plastic particles in soil: state of the knowledge on sources, occurrence and distribution, analytical methods and ecological impacts. Environ Sci: Processes Impacts 23(2):240–274

Piehl S et al (2018) Identification and quantification of macro- and microplastics on an agricultural farmland. Sci Rep 8(1):17950

Prata JC et al (2020) The importance of contamination control in airborne fibers and microplastic sampling: experiences from indoor and outdoor air sampling in Aveiro, Portugal. Mar Pollut Bull 159:111522

Primpke S et al (2018) Reference database design for the automated analysis of microplastic samples based on Fourier transform infrared (FTIR) spectroscopy. Anal Bioanal Chem 410:5131–5141

Primpke S et al (2020) Critical assessment of analytical methods for the harmonized and cost-efficient analysis of microplastics. Appl Spectrosc 74(9):1012–1047

Qi Y et al (2020) Effects of plastic mulch film residues on wheat rhizosphere and soil properties. J Hazard Mater 387:121711

Ramanayaka S, Zhang H, Semple KT (2024) Environmental fate of microplastics and common polymer additives in non-biodegradable plastic mulch applied agricultural soils. Environ Pollut 363:125249

Rillig MC, Lehmann A (2020) Microplastic in terrestrial ecosystems. Science 368(6498):1430–1431

Rillig MC et al (2019) Microplastic effects on plants. New Phytol 223(3):1066–1070

Roblin B et al (2020) Ambient atmospheric deposition of anthropogenic microfibers and microplastics on the Western periphery of Europe (Ireland). Environ Sci Technol 54(18):11100–11108

Sa'adu I, Farsang A (2023) Plastic contamination in agricultural soils: a review. Environ Sci Eur 35(1):13

Samphire M, Chadwick DR, Jones DL (2023) Biodegradable plastic mulch films increase yield and promote nitrogen use efficiency in organic horticulture. Front Agron 5:1141608

Scircle A et al (2020) Occurrence of microplastic pollution at Oyster reefs and other coastal sites in the Mississippi sound, USA: impacts of freshwater inflows from flooding. Toxics 8(2):35

Shafea L et al (2023) Microplastics in agroecosystems: a review of effects on soil biota and key soil functions. J Plant Nutr Soil Sci 186(1):5–22

Simon M, van Alst N, Vollertsen J (2018) Quantification of microplastic mass and removal rates at wastewater treatment plants applying Focal Plane Array (FPA)-based Fourier Transform Infrared (FT-IR) imaging. Water Res 142:1–9

Soltani NS, Taylor MP, Wilson SP (2021) Quantification and exposure assessment of microplastics in Australian indoor house dust. Environ Pollut 283:117064

Stanton T et al (2019) Freshwater and airborne textile fibre populations are dominated by 'natural', not microplastic, fibres. Sci Total Environ 666:377–389

Steinmetz Z et al (2016) Plastic mulching in agriculture. Trading short-term agronomic benefits for long-term soil degradation? Sci Total Environ 550:690–705

Stock F et al (2019) Sampling techniques and preparation methods for microplastic analyses in the aquatic environment–a review. TrAC Trends Anal Chem 113:84–92

Sun J et al (2019) Microplastics in wastewater treatment plants: detection, occurrence and removal. Water Res 152:21–37

Sun N et al (2023) Combined toxicity of micro/nanoplastics loaded with environmental pollutants to organisms and cells: role, effects, and mechanism. Environ Int 171:107711

Syafei A et al (2019) Microplastic pollution in the ambient air of Surabaya, Indonesia. Curr World Environ 14:290–298

Talvitie J et al (2015) Do wastewater treatment plants act as a potential point source of microplastics? Preliminary study in the coastal Gulf of Finland, Baltic Sea. Water Sci Technol 72(9):1495–1504

Talvitie J et al (2017) How well is microlitter purified from wastewater? – a detailed study on the stepwise removal of microlitter in a tertiary level wastewater treatment plant. Water Res 109:164–172

Truong TN et al (2021) Microplastic in atmospheric fallouts of a developing Southeast Asian megacity under tropical climate. Chemosphere 272:129874
Uddin S, Fowler SW, Behbehani M (2020a) An assessment of microplastic inputs into the aquatic environment from wastewater streams. Mar Pollut Bull 160:111538
Uddin S et al (2020b) Standardized protocols for microplastics determinations in environmental samples from the Gulf and marginal seas. Mar Pollut Bull 158:111374
Uddin S et al (2021) A review of microplastic distribution in sediment profiles. Mar Pollut Bull 163:111973
Uddin S et al (2022a) Microplastics in Kuwait's wastewater streams. Sustainability 14(23):15817
Uddin S et al (2022b) Micro-nano plastic in the aquatic environment: methodological problems and challenges. Animals 12(3):297
Uddin AS, Uddin S, Fowler SW (2023a) Environmental microplastics: a significant pollutant of the Anthropocene. In: Khan A, Wang C, Asiri AM (eds) Microplastic sources, fate and solution. Springer Nature Singapore, Singapore, pp 89–105
Uddin S et al (2023b) Aerosols as vectors for contaminants: a perspective based on outdoor aerosol data from Kuwait. Atmos 14(3):470
Uddin S et al (2024) Sewage sludge as soil amendment in Arid soils - a trace metal, nutrient and trace organics perspective. Emerg Contamin 11(1):100420
Uzamurera AG et al (2023) Thickness-dependent release of microplastics and phthalic acid esters from polythene and biodegradable residual films in agricultural soils and its related productivity effects. J Hazard Mater 448:130897
Vianello A et al (2019) Simulating human exposure to indoor airborne microplastics using a Breathing Thermal Manikin. Sci Rep 9(1):8670
Wang F et al (2022) Effects of microplastics on soil properties: current knowledge and future perspectives. J Hazard Mater 424:127531
Wang C et al (2024) A review of the application and impact of drip irrigation under plastic mulch in agricultural ecosystems. Agronomy 14(8):1752
Weithmann N et al (2018) Organic fertilizer as a vehicle for the entry of microplastic into the environment. Sci Adv 4(4):eaap8060
Whiting QT et al (2022) A high-throughput, automated technique for microplastics detection, quantification, and characterization in surface waters using laser direct infrared spectroscopy. Anal Bioanal Chem 414(29–30):8353–8364
WHO (2013) Guidelines for the safe use of wastewater, excreta and greywater. Excreta and greywater use in agriculture, vol 4. World Health Organization, France, pp 1–177
Wright SL, Levermore JM, Kelly FJ (2019) Raman spectral imaging for the detection of inhalable microplastics in ambient particulate matter samples. Environ Sci Technol 53(15):8947–8956
Wright SL et al (2020) Atmospheric microplastic deposition in an urban environment and an evaluation of transport. Environ Int 136:105411
Wright SL et al (2021) Development of screening criteria for microplastic particles in air and atmospheric deposition: critical review and applicability towards assessing human exposure. Microplast Nanoplast 1(1):6
Xia Y, Niu S, Yu J (2023) Microplastics as vectors of organic pollutants in aquatic environment: a review on mechanisms, numerical models, and influencing factors. Sci Total Environ 887:164008
Xu G, Lin X, Yu Y (2023a) Different effects and mechanisms of polystyrene micro- and nanoplastics on the uptake of heavy metals (Cu, Zn, Pb and Cd) by lettuce (Lactuca sativa L.). Environ Pollut 316:120656
Xu X et al (2023b) Treatment of industrial ferric sludge through a facile acid-assisted hydrothermal reaction: focusing on dry mass reduction and hydrochar recyclability performance. Sci Total Environ 869:161879
Xu J et al (2024) Quantification and identification of microplastics in organic fertilizers: the implication for the manufacture and safe application. Water Air Soil Pollut 235:169

Yongming L et al (2018) Pay attention to research on microplastic pollution in soil for prevention of ecological and food chain risks. Bull Chin Acad Sci 33(10):3
Zaman M, Uddin S (2024) Sewage sludge usage in agriculture: are we introducing microplastics in food chain? Soils Newslett, International Atomic Energy Agency 10(2):7–8
Zhang K et al (2018) Microplastic pollution in China's inland water systems: a review of findings, methods, characteristics, effects, and management. Sci Total Environ 630:1641–1653
Zhang L et al (2020) An overlooked entry pathway of microplastics into agricultural soils from application of sludge-based fertilizers. Environ Sci Technol 54(7):4248–4255
Zhang X et al (2021) Systematical review of interactions between microplastics and microorganisms in the soil environment. J Hazard Mater 418:126288
Zhang S et al (2022) Occurrence and distribution of microplastics in organic fertilizers in China. Sci Total Environ 844:157061
Zhang S et al (2024) From organic fertilizer to the soils: what happens to the microplastics? A critical review. Sci Total Environ 919:170217
Zhaorong M et al (2020) The usage and environmental pollution of agricultural plastic film. Asian J Ecotoxicol 4:21–32
Zhou Q, Tian C, Luo Y (2017) Various forms and deposition fluxes of microplastics identified in the coastal urban atmosphere. Chin Sci Bull 62:3902–3909
Zhou B et al (2020) Microplastics in agricultural soils on the coastal plain of Hangzhou Bay, east China: multiple sources other than plastic mulching film. J Hazard Mater 388:121814
Ziajahromi S et al (2017) Wastewater treatment plants as a pathway for microplastics: development of a new approach to sample wastewater-based microplastics. Water Res 112:93–99

MIX
Papier aus verantwortungsvollen Quellen
Paper from responsible sources
FSC® C105338

If you have any concerns about our products,
you can contact us on
ProductSafety@springernature.com

In case Publisher is established outside the EU,
the EU authorized representative is:
Springer Nature Customer Service Center GmbH
Europaplatz 3, 69115 Heidelberg, Germany

Printed by Libri Plureos GmbH
in Hamburg, Germany